# About McLachlan's other books

*With Unshakeable Persistence: Rural Teachers of the Depression Era*

"These stories will inspire today's teachers and practicing classroom educators in their own modern day efforts to provide the best possible learning environments in today's schools—all of which face problems unique to our own troubled times."

—*Wisconsin Bookwatch*

". . . describes the humorous and sometimes tragic experiences the teachers faced . . . a treasure of memories that are an integral part of Alberta's heritage, and a wonderful window into the past."

—*Alberta Views*

*With Unfailing Dedication: Rural Teachers in the War Years*

"Through the first-hand recollections of the teachers, these stories add a valuable contribution not only to Prairie history, but to Canadian education as well."

—*Prairie Books Now*

"This book gives us rare insights into the lives of these stalwart people who taught in our schools during and after the Second World War. . . . It is another triumph for McLachlan."

—*The ATA News*

# Gone But Not Forgotten
## Tales of the Disappearing Grain Elevators

Salvador, SK
PHOTOGRAPHER/CONTRIBUTOR: CHRIS STACKHOUSE

# Gone But Not Forgotten
## Tales of the Disappearing Grain Elevators

ELIZABETH McLACHLAN

INTRODUCTION BY SHARON BUTALA

NEWEST
PRESS

**National Library of Canada Cataloguing in Publication**
McLachlan, Elizabeth, 1957-
Gone but not forgotten : tales of the disappearing grain
elevators / Elizabeth McLachlan.

ISBN 1-896300-76-6

1. Grain elevators—Prairie Provinces—History. 2. Grain
elevators—Prairie Provinces—Pictorial works. I. Title.

HD9044.C23P73 2004     633.1'0468     C2004-900768-8


Editor for the Press: Lynne Van Luven
Front cover photo: Chris Stackhouse, www.chrisstackhouse.com
Back cover photo: Teena Feniak
Cover and book design: Ruth Linka
Copyeditor: Christine Savage
Interior photos: Photographer and contributor are listed with each photo.

NeWest Press acknowledges the support of the Canada Council for the Arts, The Alberta Foundation for the Arts, and the Edmonton Arts Council for our publishing program. We also acknowledge the financial support of the Government of Canada through the Book Publishing Industry Development Program (BPIDP) for our publishing activities.

NeWest Press
201-8540-109 Street
Edmonton, Alberta
T6G 1E6
(780) 432-9427
www.newestpress.com

1   2   3   4   5   07   06   05   04

PRINTED AND BOUND IN CANADA

THIS BOOK IS PRINTED ON ANCIENT-FOREST-FRIENDLY PAPER

*To those whose lives were shaped and sustained
by the prairie elevators of our past.*

There's a quota on! Trucks lined up to the elevators.
Valley Centre, SK, 1950
PHOTOGRAPHER: UNKNOWN / CONTRIBUTOR: VERLA NEVAY

1. Driveway and receiving scale
2. Grain pit
3. Elevator 'leg'
4. Distributing spout
5. Spouts to bin
6. Loading spouts to boxcar
7. Return spout to truck
8. Storage bins
9. Scale hopper
10. Shipping scale
11. Grain cleaner
12. Spout to truck
13. Spout from cleaner to scale
14. Driving motor
15. Cleaner bin

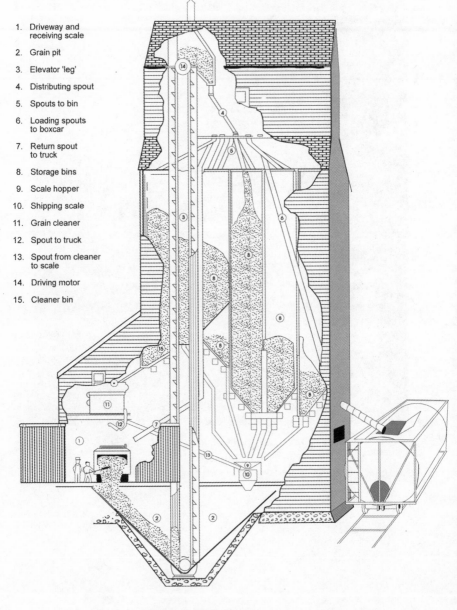

IMAGE COURTESY OF THE CANADIAN INTERNATIONAL GRAINS INSTITUTE
WWW.CIGI.CA

# TABLE OF CONTENTS

PHOTOGRAPHER/CONTRIBUTOR: STEWART GRAHAM

# Absences

BY SHARON BUTALA

*Gone But Not Forgotten: Tales of the Disappearing Grain Elevators* by Elizabeth McLachlan is a new approach to a familiar subject. Grain elevators were, for prairie people, more than merely a place to store grain. They were a symbol, too, not just of a way to make a living, but of an entire way of life. This book tells how elevators were an integral part of the prairie experience.

One of my girlfriends used to support her career as a graduate student by creating batik wall hangings and selling them at craft fairs. The year she hit on the idea of using grain elevators as a subject, she cleaned up, making an unforgettable thousand dollars in one day. I can hardly believe it, but that was thirty years ago, and she has been a respected professor for a long time. Grain elevators, in the meantime, as this book will tell you, have had less good fortune. In fact, although I didn't know it, the golden age of wooden elevators had already peaked before I was born, around 1934, when there were about six thousand of them dotting the prairie landscape. They were everywhere, their silhouettes on the horizon the first sign of the welcome presence of people. In fact, they were such a commonplace that people of my generation no longer paid much attention to them. They were just there; they were just life. I know that by the time I was an adult I had long since stopped seeing them, even when I was staring right at them.

When I was a child, I saw grain elevators. They seemed beautiful and mysterious to me, the Notre Dame Cathedrals, the Edinburgh Castles of our prairie villages, always by far the tallest buildings and bigger even than the schools or the churches. They seemed as inevitable and right as the wide fields of grain around them. I'm pretty sure that I never even questioned where they'd come from or what they were for, although probably I had only the vaguest notion, since by the time I was born, our father was making his living as a mechanic and our mother was a housewife, both of them having long since been driven from the family farms where they were raised, and the farms themselves long owned by strangers.

But every fall when we reached the railroad tracks we had to cross on our way to school, we would be excited to find a half-mile-long line of trucks parked down the side of the street, each with a single driver, each truck groaning with its heaping load of grain, and slowly, ever so slowly—parked much longer than it moved—eventually snaking its way into the elevator where its load would be weighed, evaluated as to grade and class, and dumped. The drivers sat alone high up in the truck cabs, looking bored, eyeing us without much interest, or standing in small groups in front of one of the trucks, chatting about what we never knew. It was all part of the mysterious world of grown-ups, just something we accepted as inevitable, like the fact that when fall came, like it or not, we all had to trek off to school. I know I assumed that, as with all things in the small-town prairie world I lived in, elevators had always been there and always would be.

For more than eighty years the wooden grain elevator had been the most telling symbol of the Canadian prairies, partly because it was our only truly indigenous architecture after the sod shack, and because it was so striking a vertical presence on the vast, empty, and sometimes flat plain. The simplicity of its design was exactly right in the midst of the stark sweep of land and enormous sky, so visually exciting that artists often used to render it in watercolours, and amateur photographers visiting from elsewhere never failed to take a picture of it, not least because the name of the town was always painted high on its wall, and in huge letters that could be seen from a long distance away. Usually, too, there were several of them in a row, one belonging to each of the competing grain companies. The bigger the town, the more of them there would be.

In some ways the elevators were tyrants. They were set where they were by the railroad companies that determined they would be about ten miles apart, this because a team pulling a wagonload of grain could make only a twenty-mile round trip in one day. And the railroad companies named the towns, most often with names that had nothing to do with physical features or history of the prairies, as places had always been named by First Nations people, and as had the ancient towns and villages in Europe. Besides that, often the elevators were built at points that were not really places, in the sense of being perhaps an attractive, wooded curve of a stream or river, or a pleasant, sheltered band of hills. They were set out on the bald prairie, and little cafés, beer parlours, stores, churches, and schools grew up around them.

But one day, elevators began to disappear. A town that once had five now had only three, or two, or maybe only one huge gold and orange one with 'Pioneer' painted on it. And great grain 'terminals' made of concrete, without beauty or mystery, signifying only the industrialization of agriculture, began to appear by the side of major highways. The old "ten miles to the nearest elevator" was becoming a thing of the past, and everywhere farmers grumbled about the long and expensive thirty- and forty-mile drives they now had to make with their grain. The villages themselves began to disappear too, with the disappearance of the elevators, if they had no other reason—as many of them did not—to exist.

Claydon, the once-lively village nearest the ranch, the one to which my husband used to take our grain, still had an elevator though, and all the nearby farmers were glad of it. One day, a woman living in a farmhouse a couple of miles from it happened to be looking out her window during a thunderstorm and saw lightning strike the elevator, and the fire that would destroy it begin. Several days later the elevator was gone and there was only a pile of blackened, burned grain still smouldering where it had once been. Farmers stood around watching the smoke rise from it, their half-tons parked haphazardly behind them along the gravel road, and kicked the dust with their boots, as they muttered about what would happen now.

A few days after that, they held a worried meeting in the nearby hall, everyone knowing that no grain-buying company would ever build another elevator in such a remote place. A temporary solution was found, but everyone knows that's what it is, temporary. Any day now there'll be no choice but to drive fifty or seventy-five miles with each load. But although the loss of the elevator was calamitous, at least, as happened in many other communities, it did not signal the death of a town. By the time that happened the village was already dead, having only a tiny post office left, an occasionally-used community hall, and a row of empty, falling-down houses in which no one had lived for a long time. There was a feeling of inevitability about that loss, I think, among all of us. After all, elevators were going down everywhere across the prairies, and we were far from major highways or cities; we all knew that one day soon the elevator would go, if not in the way it did. There was an irony about that particular loss that everyone felt, but on which no one commented aloud.

But driving down the road toward the ranch a week or so later, forgetting about the fire, I was for a minute lost, disoriented, the one landmark that had always signalled my near-arrival gone, leaving only a blank space on the low horizon. Coming upon the place suddenly I was unprepared, and the unexpected emptiness of that windswept, grassy spot struck a plangent chord in me of loss, the absence of that elevator having now become as powerful as its presence had been.

Sharon Butala
Eastend, Saskatchewan
December 2003

The flames ate their way steadily through the cribbed
two-by-fours. Granum, AB, 1982

PHOTOGRAPHER/CONTRIBUTOR: ELIZABETH MCLACHLAN

The baby was finally asleep. The lunch dishes were cleared, and my husband had returned to work, trudging up frozen alleys to the Alberta Wheat Pool elevator. Bundled in layers of sweaters, I sat cross-legged on our rickety bed, labouring with numbed fingers over a letter to my parents. The lady who controlled the thermostat in our motel-style eightplex lived next to the boiler room. It was hard to get her to turn up the heat. The cinder block walls and thin carpet did a better job of keeping the cold in than out. So did the door that opened directly off the living room onto the street. The mid-January freeze was so deep that even Granum's water system had succumbed. While crews struggled to fix it, the town's four hundred residents had water for only an hour a day. Between 7 and 8 AM, I washed diapers. For the twenty-three hours following, the taps yielded not a drop.

Nothing around me was familiar. All our belongings were in storage while Dale finished his last three months of the Alberta Wheat Pool Management Training Program in this small Alberta town forty miles from Lethbridge. In six weeks he'd be an Assistant Elevator Manager, and we'd be moving again. My pen rasped across the page. The clock ticked quietly. My eyelids drooped.

1

Suddenly the flimsy door crashed open. Dale burst through, a whirlwind of flailing arms and churning feet.

"The elevator blew up!"

"What?"

"It exploded! Where's the phone book?"

I looked out the window. Across the empty street and down the alley, the elevator stood majestic and silent, just as it always had.

"It looks fine to me."

"It's on fire, I tell you! I have to call Harry."

Harry Harsula, the travelling supervisor for the Granum region, was the man who'd hired Dale. He also had the power to fire him.

"Damn!" Dale was fumbling with the phone book. "The manager away. Just me and Gerald . . . I don't know what happened. We were elevating grain and suddenly the leg quit. A great honkin' piece of charred belt fell right out of it."

"What about the fire department?"

"We called them. Then we had to get everything the hell out. Those permit books . . . I didn't think my legs would get me here. I was running and running, but they weren't moving. Oh God! The town's got no water!"

He made his quick phone call and was out the door, leaving the baby still asleep and me standing bewildered, staring at what still appeared to be a perfectly intact elevator.

Within thirty minutes, however, a line of orange crept down between the elevator and its annex. The flames blistered the green paint, curled around the corners of the building, and ate their way steadily through the cribbed two-by-fours, tinder-dry and stacked like layers of a sandwich. A growing crowd of bystanders could only watch, fascinated, as the elevator was slowly devoured.

Fire trucks rushed from surrounding communities to support Granum's meager department, but they were little more than a bucket brigade as their tanks quickly emptied. For the rest of the night, farmers hauled water while firemen turned their backs on the inferno and strove to save the town, soaking buildings in every direction, including those on main street across from elevator row.

For days afterward, mountains of grain smoldered and stank. There was an investigation, of course. With the manager on holidays in Las Vegas, Dale and the assistant manager were accountable. They insisted they'd done nothing wrong. But they were guilty until proven innocent. Everybody knew that elevators were giant matchsticks. Little more than a spark could ignite them. Once kindled, they'd been known to set alight entire towns. The company strictly enforced fire safety procedures, including nightly inspections. Anyone found shirking was quickly out of a job. Gerald and Dale had been diligent, but still the elevator burned. Gerald, as the one in charge, was on the hot seat for days. Dale, just a rookie, envisioned his career going up in smoke. In the end, they were exonerated, but the experience was devastating. The harm that one burnt-out motor bearing can do!

◆◆◆

Twenty years later concrete "super elevators" are replacing wooden ones at a stunning rate. They're not nearly the fire hazard their predecessors were, but neither do they hold the charm.

"The huge cement structures that are built now for grain deliveries don't excite me," says Verla Nevay, a former grain buyer's wife. "They are gray, cold looking, and mysterious."

However, their coming of age has been decreed by our super-sized society: larger farms, semi-sized trucks, rail line initiatives to load more at each point, closure of small branch lines, and the merging of grain companies into conglomerates.

"I always felt I had a good day if I loaded four cars a day," reflects Gerald Hallick, who ran the Ogilvie elevator in Rosser, Manitoba, from 1948 to 1968. "I understand that they can now load those new cars in eight minutes."

In fact, new concrete elevators boast ten times greater speed and handling capacity. It's no wonder each high throughput facility that goes up foreshadows the same number of wooden elevators coming down. There is an irony here. The same progress that is destroying the wooden grain elevators today is what brought them to the prairies in the first place.

Leonard Wold of Alliance, Alberta, recalls how grain was handled in the early years of Western Canadian settlement.

"In about 1908 Grandpa hauled wheat by horse and wagon to Sedgewick, as the rail line only extended that far. There were no elevators to elevate grain yet, so each farmer shovelled his wheat into the railroad cars, which were wooden at that time."

"Down the track a short distance from the train station was a loading platform," adds Claude Stevens of Foremost, Alberta. "The boxcar floor was even with the platform. Farmers loaded their own boxcar directly from their wagon."

Alternatively, grain was poured into one hundred twenty-pound (two bushel) sacks straight from the threshing machine, then stacked in wagons for the trip to the nearest rail line. It took about six hundred fifty sacks to load a boxcar.

"Back when men were men," says Claude, "you walked into the boxcar with a sackful on your back and piled them up to shoulder height at both ends and a few at the doors."

Soon more settlers and improved farming methods resulted in additional acres of cultivated land and a dramatic increase in grain production. The need for better storage facilities at railroad sidings was crucial. Traditional flat warehouses were proving impractical and inadequate. Elevator models, which functioned on the principle of gravity, had already been experimented with for several years. The first, a squat, round structure, appeared in Niverville, Manitoba, in 1879. It was built by A.E. Hespler, an immigration agent responsible for bringing sixty-five Mennonite families into the region five years earlier. Two years later, in Gretna, Manitoba, the Ogilvie Milling Company erected a square model, after which future elevators were patterned. By the turn of the century, elevators outnumbered warehouses.

As the Canadian Pacific Railway recognized the vast potential of the west, it encouraged the construction of even more elevators. Roughly every ten miles along its lines, the railway offered free land to anyone willing to raise an elevator. Ten miles was considered a comfortable distance to haul a load of grain by team and wagon in one day. The rules were simple: the building must be equipped with a steam or gasoline engine, a grain cleaner, and the capacity to store a minimum twenty-five thousand bushels of grain. By 1913 a standard elevator emerged. Run on a fifteen-horsepower engine, it measured thirty-one by thirty-three feet, and rose as high as eighty feet into the air. It held thirty-five thousand bushels of grain in sixteen to eighteen bins. Non-standard structures continued to be built, but those of standard design far outnumbered them. Thus, the familiar house-shaped tower with crowning cupola was born.

By the 1933–34 crop year, almost six thousand elevators dotted the prairies, firmly establishing their long-standing identity as symbols of progress and prosperity. Long-standing, but not permanent.

In the beginning they were primarily owned by small groups or individuals, but as the advantages of consolidation became clear, owners began to string together to form small companies or lines. There were over three hundred in 1916. By 1985, however, a mere six companies owned over 95 per cent of prairie elevators. In 1997 only eight main-player companies remained. As consolidation continues, the number of elevators declines. By December 2002 only 419 remained in Canada. They continue to fall like dominoes. Claude Stevens points out the paradox of losing, in the name of progress, local wooden elevators to major concrete terminals.

"When Dad hit the country in 1909 he hauled grain forty miles to Grassy Lake until the railway was built to Foremost in 1913. Now, 2002, instead of three miles to Foremost we have to haul to Stirling, fifty miles, Bow Island, thirty miles, or back to Grassy Lake, forty miles."

Elevators don't just represent progress. They are monuments to prairie life. Prairie people gave them breath and meaning. People built them, ran them, relied upon them, lived in them, and died in them. Our temptation to romanticize a fading era is almost irresistible, but strip it away and what have you got? You have Leonard, Verla, Gerald, Claude, Dale, and all the others in this book—real people with real stories. They tell of hard work and hope, of humour and heartbreak, and above all, of solemn affiliation to the land. These pages tell their stories—an unadulterated portrait of prairie life.

# CHAPTER ONE

# "You Stupid Idiot!"

"I don't remember my brothers ever hauling grain," chuckles Clarence Cluff. "They somehow got out of it. But I kind of liked it. I liked it in the winter better because sleighing was smoother than the wagon."

Living just a few miles from the Cluff homestead near Kyle, Saskatchewan, John Hirtz might have disagreed.

"The winter of 1916 John . . . hauled grain from south of Beechy to Herbert," writes his granddaughter Val Hvidston eighty-seven years later. She has written John's story based upon his journals. "That winter was so cold, many times he froze his face, feet, and hands. Once so badly that John's face blistered and did not heal up until spring. He stopped overnight at a farmstead near Main Centre, and he wished he had stayed in Luxembourg."

Even Clarence concedes that "sometimes in the wintertime it was cold." Nonetheless, the grain had to get to market, and for Clarence and his dad, that meant an overnight journey.

"When we hauled to Success we had to ferry across the South Saskatchewan River. In the winter we drove across as soon as the

Horse-drawn grain wagon. Pre-1926
PHOTOGRAPHER: UNKNOWN / CONTRIBUTOR: MARIE BARNEC

river ice would hold. Or we could go the other way to Elrose. One was thirty miles; one was twenty-eight. If we hauled to Success, we had a little better freight rate because it was on the main line. I think it was a matter of five cents a bushel or something, but that was a lot of money in the 1920s and 1930s. We needed it."

Clarence shovelled wheat from the bin with a grain scoop.

"It was kind of a pail that had two handles. It held half a bushel. You always tried to go without stopping. That meant you had to throw all the way to the front of the wagon. I could because I could go right- or left-handed, but that's the only pause Dad allowed, to change hands. 'You can't stop,' he said. 'You've got to keep going!'"

Clarence was tall, lean, and strong. He soon learned to handle the job with ease. "One hundred and twenty scoops were sixty bushels. When I came to one hundred, I knew I only had ten more bushels to go."

The two-decker wagon took approximately sixty bushels and the three-decker took eighty. Larger, curvier grain tanks took up to one hundred twenty-five, but the Cluffs used the smaller outfits. By 7 AM they'd hitched a two-horse team to the wagon and were making tracks down the dirt road to Success.

"We had to have good horses because we had to go downhill to

Success, SK
PHOTOGRAPHER/CONTRIBUTOR: CHRIS STACKHOUSE

the ferry, and the horses had to hold that load back. We had brakes on the wagon but we were always reluctant to use them because we wanted to make it as easy for the horses as possible. We had to put the brakes on for the big load (three-decker wagon). No horse could hold that back with just their neck yoke and martingale."

They met buggies, wagons, and the occasional motorized vehicle on the narrow trail. As a rule, loaded wagons had the right-of-way. Anyone meeting them was expected to leave the path and bypass them on the open prairie.

"Once you got out of the valley, the rest of the road was fairly good," says Clarence. "I wouldn't say it was two-way, but it was in a way. There were lay-bys, places where you could pull off and wait. Whoever was closest to it had to pull over. Once you got past it, if the other guy didn't stop you were in trouble, but everybody was pretty good."

The trip took eight hours, leaving Clarence and his dad enough daylight to deliver the grain to the elevator and reload their wagon with coal or other supplies for the trip home the next morning. At the end of a long day, they turned the horses into the livery barn and headed to the local hotel for the night.

"The restaurant usually opened around 6 AM," recalls Clarence. So by 7 AM they were on their way home again.

"In a good year, when we had lots of grain, we did this twice a week sometimes."

Soon afterward the Canadian Pacific Railway Company extended its line to the Cluff farmland.

"They took thirty-five acres of our land," says Clarence, "and they built a Y to turn the trains around. Sometimes the trains stayed overnight. They had a cabin for the engineer and the fireman."

Before long, elevators were raised on the site and the town of Matador slowly emerged around them. By the time he was sixteen, Clarence found it natural to help out in the Matador State Elevator during busy seasons.

"The elevator handled coal. Sometimes I'd get a job unloading a car of it in the winter. If the farmers came in, I could load them right out of the car. Otherwise, I had to have a wheelbarrow to get it to the coal shed. Fifty cents a ton is what I was paid. Forty tons in a car times fifty cents was twenty dollars."

Being suddenly on the receiving end of grain deliveries was an eye-opener for Clarence. Those were the days of hand-cranked wagon hoists. Later they were replaced by air hoists, but the procedure was the same. Farmers pulled their wagons up the ramp and through the large doors of the lean-to-style drive shed. Once inside, they guided their horses over the grated pit and onto the hoist of the large scale. Then they hopped down by way of the front wagon wheel. Lloyd Henders of Austin, Manitoba, was thirteen when he hauled grain from his family's farm in the Culross–Elm Creek area of Southern Manitoba. He marvels at the discipline of the horses.

"As I guided that team up the incline to the elevator, I prayed that they would not spook at the sudden thunder of their hooves on the planks and steel grating over the pit. The next challenge was to get them to stop with the front wagon wheels placed precisely on the steel platform of the hoist."

The elevator operator weighed the wagon, marking the weight on a scale ticket. Then he opened the end gate. If using a crank hoist, he unlocked a wheel-and-chain mechanism attached to the side of the scale and turned it to raise the front of the wagon. Its rear dropped simultaneously and grain flowed into the grated pit beneath the driveway floor. With an air hoist, such as Lloyd was on, cranking was unnecessary. The agent simply pulled a lever and let air pressure do the work.

"The horses were left hitched to the wagon during the whole cycle," says Lloyd. "They stood there while the tugs and centre pole rose almost to the level of their backs and then were lowered. And all this time there were the usual noises and dust of the elevator swirling around them. If one of those animals had panicked at the wrong moment, there would have been serious damage and possible injury."

"We liked to get the wagon balanced on the scale perfectly," says Clarence. "If it was, we wouldn't have to do any work at all. Once it started to dump, it went fast. It took about three minutes if the pit was empty."

As the grain flowed, the elevator agent caught a sample in a tin pan or scoop. By determining the grade (quality) of the sample, he knew what to pay the farmer per bushel for his load. Once the wagon emptied, it was righted and weighed again to calculate the number of bushels delivered.

At this point the farmer was free to drive or lead his team out of the exit doors of the drive shed.

"One might speculate on why I was so fortunate to have such well-mannered horses," says Lloyd. "Maybe they had it figured out. Once that wagon came off the hoist it pulled a lot easier, and they could go home for oats. They may also have expressed another opinion because, after they left the elevator, the agent often had to use his scoop shovel for something besides grain!"

Of course there were the occasional wily farmers who knew how to take advantage of the horses remaining hitched to the wagon.

"There were men who'd stand with their team," says Clarence, "and when they saw the elevator man weighing, they'd pull back on the lines. With the horses backing up it would put a little weight on the wagon. It didn't move the wheels; it just pushed a little weight on them. They could get up to thirty more pounds on the scale. Some of those farmers! Their teams were so well trained they'd just lean back a little automatically."

Elevator operators usually figured out who the offenders were. It got to be a joke.

"Did you give so-and-so an extra bushel?" they'd ask each other when they met.

Clarence didn't mind helping in the elevator.

"I rather liked it in a way. I felt kind of important as a sixteen- or seventeen-year-old manning the old hand crank. The pit was an immense affair, same as it is now, because the back uses it too. The front (driveway) end only used half the pit because the leg was in the middle."

The leg was an endless belt, about eighteen inches wide, fitted with cups and extending in a closed shaft from the pit to the top of the elevator. Once grain was delivered to the front pit, Clarence started the one-cylinder gas engine that powered the leg. It scooped

Rope-driven leg with inspection door open. Note the cups inside, the bin wheel above the door, and the grain shovel resting against the wall. In the foreground, the scale platform and grating over the pit can just be seen.
PHOTOGRAPHER/CONTRIBUTOR: CHRIS STACKHOUSE

the grain up to a distributor at the top of the elevator. From there it was directed into bins to await loading into boxcars.

One harvest night, due to the abundant crops, the elevator was almost full.

"Farmers were just hauling in from the thrashers," says Clarence. "They were lined up as far back as you could see. The CPR left five boxcars for loading and rather than elevating grain into the bins we were just running it right through to the car. It was awkward trying to load a car and receive grain at the same time. You could see the farmers were gaining on us. But we knew it was all going to be over in a week or so."

Clarence was trackside, up on a ladder against a boxcar, directing grain through a loading spout. All he had for light was a kerosene lantern.

"I held it in the car once in a while to see if everything was all right, but most of the time I knew what I was doing in the dark."

In the driveway the elevator agent too had only a kerosene lamp. It was meager light by today's standards, but enough for the horses to see their way onto the scale. One by one they took their places and patiently waited for the loads they pulled to be weighed and dumped.

Everything ran like clockwork, until suddenly the agent took a notion to check on Clarence. Without thinking, he picked up the kerosene lamp and made his way down the narrow corridor to the other side of the elevator. The farmer next in line was just driving his team up the ramp. Instantly he was cloaked in darkness.

"He couldn't see to steer," says Clarence, "and you can't stop a team when they're pulling uphill. He had to depend on the horses to know where to go, but they were off the scale. They hit the crank with the hub of the wagon wheel."

A loud crash splintered the air, sending the agent rushing back with the lantern. He was shocked to discover the hoist in pieces.

"It looked like a disaster," says Clarence. "The farmer broke it right off."

The agent was furious, but he couldn't deny that he'd been at fault by taking away the light. He was certain they'd lost the use of the hoist and from then on would have to shovel every wagon off by hand, putting them even farther behind at the busiest time of year.

Clarence took a close look at the wreck and informed the agent that they could probably fix it.

The agent didn't believe him, but Clarence was determined.

"Nothing metal was broken. The chain was twisted but it wasn't out of action. The elevator had just been built, so there was still lots of lumber around. I ran home and got a brace and bits and a couple of wrenches. I brought in a couple of lanterns and a sawhorse. The chap who ran into it was a bit of a carpenter and he helped. Even though it was in the dark and all we had was lanterns to work with, in less than an hour we had it up."

In the meantime the agent set the leg to fill the only bin with space remaining, and he and the waiting farmers kept the line going, shovelling grain by hand.

"I think they unloaded three wagons," remembers Clarence. "We had all the rest dumped by midnight. I got the boxcar loaded and the next morning I got another couple of them loaded and then the train came, hauled them away, and brought some more."

The crank hoist repair was only supposed to be temporary, but Clarence's talent for fixing things had surfaced. The hoist functioned admirably until there was finally a lull in harvest activity.

When he grew older, Clarence went to work for a farmer near Semans, Saskatchewan. Compared with his earlier experiences hauling grain to Success, the four-mile jaunt to Semans with a one hundred twenty-five-bushel tank was a breeze. Clarence always travelled sitting on top of the load.

"There was a special wagon seat that a lot of people had," he comments. "It had springs on it. Oh, if you were hauling thirty or forty miles it was all right, but ten miles, sitting on the grain was just as good. There was lots of room."

One unfortunate day Clarence could have done with a proper seat.

"I had four horses on the empty grain tank and I was going home. A terrible thunderstorm blew in. It was such a downpour. The lightning was right along the wire fence. You could see it. I got about a mile down the road before the horses took off."

Hell-bent for who knows where, the spooked horses were out of control.

"They were kicking mud, the front ones kicking mud on the back ones. I was hanging on to them, trying to keep them straight—and oh, I'm thankful I never met anybody. All of a sudden, I looked down and I was standing in water."

The downpour was quickly filling the grain tank. Clarence stepped up onto its flared sides, awkwardly straddling the water while he struggled to master the frightened horses.

"I was trying to drive the horses, trying to keep out of this damn water, and I saw a straight stretch coming. So I tied the horses' lines around the stake, worked my way back, reached over and opened the end gate."

A flood of water gushed from the grain tank. Clarence left the gate open and turned back to the horses.

"When I got home I told the owner about it. 'You stupid idiot!' he said. 'You should have left the tank full. It would've slowed the horses down.' I was cutting up the road something terrible. I figured I was going to get shot. What a mess the horses were. I was too, but I didn't dare quit until I got them cleaned up."

Eventually Clarence made his way to Alberta, where he farmed and raised Black Angus cattle for many years. Now ninety years old, he is still lean and strong, with the stride of a man half his age.

"I feel like I'm eighteen," he says.

To keep his mind sharp, he doesn't just play bridge—he teaches it. Although retired, he often works six hours a day restoring artifacts at the Coaldale Gem of the West Museum. His specialty is repairing old farm equipment.

# CHAPTER TWO

# Runaway

Young Boris Johnson gazed in dismay as the empty manlift, with a great rush and clatter, disappeared into the upper regions of his father's elevator. He had planned to be on it, but the darn thing got away on him. He didn't have permission to use it in the first place. Father would be livid.

Those manlifts were tricky and even dangerous. Operated on a system of weights and counterweights, the primitive lifts were designed to transport personnel to the top of the elevator quickly. If things had gone properly, Boris would have adjusted the counterweight for his eighty-pound build, let off the safety lever, released the foot brake, then used the rope to pull himself to the top of the building in a controlled manner. But somehow he'd released the safety without stepping onto the platform to provide weight, the brake had failed, and without warning the manlift had taken off without him at lightning speed, crashing full bore into the beams above.

Elevator staff travelled the manlift regularly for several reasons. Grain that moved up the leg to the top of the elevator was distributed to the proper bin via a spout, connected by a cable to a casino-like wheel on the driveway floor below. The operator spun the wheel to align an arrow with a bin number, and, way up above, the spout rotated to the corresponding bin. Because the operator couldn't see what was happening at the top, he frequently rode the manlift to make sure everything was running smoothly. This was wise. While helping his father, Boris once almost mixed barley with wheat. Such mistakes were not uncommon.

"If one worked long enough in an elevator, the day would come when a load of grain was dumped into the wrong bin," says Kevin Marken, a former agent for Pioneer Grain in Schuler, Alberta. "The worst was when one type of grain got dumped on another. I'd have to drain the grain down in the next bin to the approximate level of where the mixture started, then drill holes in the wooden crib wall until I found the level where the two grains met. Once I found it I'd cut a larger hole in order to drain off as much of the grain as I could before it got really mixed. It was a long, hard process, so it was well worth the extra trip up the manlift to verify that the spout was set over the correct bin. I think the helper that I had one year was getting frustrated with me when I was always asking him to go upstairs to verify that the grain was going exactly where it should be going!"

Elevator operators developed a sixth sense for the wheel. With surgeon-like skill, they knew when something didn't "feel" right. Usually it meant the spout hadn't slipped properly into the saucer at the opening of the bin, or that it had stopped between two bins. A trip up the manlift to make adjustments was necessary. The manlift also provided access to the cleaner floor, bin floor, and cupola floor. These landings on the way to the top needed to be regularly swept free of the ever-present grain dust, which is incredibly explosive in high concentrations. In later years, when electricity was installed, the motor in the cupola had to be checked every day to ensure that no hot bearings created a fire hazard. A system of ladders climbed eighty feet to the top of the elevator, but the manlift was quicker and far easier.

It was to these ladders that a frowning Gilbert Johnson now steered his son. They were thick with dust. Boris would be covered by the time he reached the top. But someone had to retrieve the runaway manlift, and he was responsible.

Occasional mishaps aside, the young boy, barely into his teens, was a great help to Gilbert, who ran the Saskatchewan Wheat Pool elevator at Marchwell in the 1930s and 1940s. One of Boris's biggest jobs was preparing boxcars for loading. When the train spotted them to the elevator, they had to be cleaned of any residue left from whatever had been shipped last.

"Imagine opening a boxcar that had been standing in the sun all day and finding that its previous cargo had been coal," says Art Braun of Altona, Manitoba, who, like Boris, also helped as a boy in his father's elevator. "Not only was it a daunting task to sweep clean, but in the 1930s hobos riding the rails would probably have used that car as a toilet."

Ken Warriner, former district manager for Federal Grain, recalls an agent from Mundare, Alberta, who opened a boxcar door one morning to find a dead body inside.

"He wrapped him up in a tarp," says Ken, "and put him in the coal bin. It was three or four days before the police came. Good thing it was cold weather. Hobos would get into the boxcar and with all the shunting, the door would get closed and then they couldn't get out. They were trapped."

"There was always grain left in the cars," remembers Gerald Hallick, "and most of the time it wasn't the kind you were loading. There were old broken grain doors and once I got a car that had one of those old two thousand pound safes in it. I don't know if there was. any money in it, but I didn't waste my time checking it out."

A hopper through which grain was weighed before being dropped into the back pit in preparation for loading into boxcars. Note the balance beam scale on the front of the hopper.

PHOTOGRAPHER/CONTRIBUTOR: CHRIS STACKHOUSE

Regardless of what was discovered in the cars, they had to be cleaned before loading. The shunting Ken mentions often bashed holes into the ends. Boris patched these with heavy paper or bits of wood and tin before the car was coopered. Coopering involved nailing grain doors over the openings on both sides of the car to create a closed container. Early grain doors were wooden. Later they were heavy cardboard reinforced with iron bands. If flax was to be shipped, the interior of the car was also lined with thick brown paper to prevent the fine, flat kernels from slipping through the cracks. The grain door on the side adjacent to the elevator didn't quite reach the top, allowing space for the loading spout to enter the car.

Once Boris finished coopering, his father took over. He extended the flexible loading spout from the elevator into the boxcar. Then, by pulling the right lever, he opened the bin holding the grain to be shipped and dropped a carload's worth into the back pit. He started the leg and elevated the measured grain up to the distributor, where it was sent rushing down the loading spout.

Dispersing grain evenly into the car took practice. With the help of a rope on a pulley, he guided the spout to aim first into one end of the car and then into the other. Some cars held a thousand bushels, others sixteen hundred. The trick was to aim most of the grain into the corners so that it wouldn't pile up in the middle and spill over the top of the grain door. Once the car was loaded, a smaller grain door was fitted into the gap where the loading spout had been. The sliding boxcar door was closed and a metal tag inserted through the handle to seal it. It wasn't much security, but printed clearly on the car was the admonition that to break the seal was to break the law and risk a heavy fine.

Boris developed a lot of muscle helping his father move loaded cars down the track and bringing the next one in line with the loading spout. A full boxcar weighed up to 160,000 pounds. Boris describes the tool they used to move it as "a winch-like device, similar to a crowbar with a long wooden handle." First, they climbed up the car to release the brake by pressing a long handle. Then they climbed down and jammed the device under one rear wheel. By pushing and pumping downwards as hard as they could, they succeeded in moving the wheel a fraction. That was usually all that was needed. Tracks beside the elevator were supposed to be laid at a gradient that allowed the boxcar to keep moving once it started. The next challenge was getting it stopped before it gathered too much speed and went too far. That required chasing it down and climbing back up to the top to crank a wheel that tightened the brake once more. Or, not recommended but often done, letting the car crash to a halt when it hit the braked car previously loaded.

There were other methods of moving cars too. In many elevators a local farmer was sometimes called upon to bring in his tractor. By running a cable from the boxcar to the tractor hitch, he made quick work of pulling cars down the track. It was a terrific help to the agent, but Lloyd Henders of Austin, Manitoba, remembers at least once when such an operation went drastically wrong.

"During dry falls, when threshing was in full swing, elevator operators tried to hire enough help to run twenty-four-hour days. With only one elevator leg, they would receive grain during the day and then switch over to loading boxcars all night. At one particular station a local farmer with an old, lug-wheeled tractor was hired to assist the elevator workers in moving cars during the nighttime loading. One night, with only kerosene lanterns for light, and amidst the engine noises and swishing grain dust, there was a fatal miscalculation. The tractor operator apparently stopped his machine, but the momentum of the loaded boxcar carried it past the stationary tractor. The towing cable now began to pull the other way. It upset the tractor on top of its operator and killed him."

This was a shocking tragedy, particularly during the 1930s when a farm widow suffered twofold, crippled both by the depression and the loss of her husband.

But in Marchwell, Boris had no such worries. He was just a boy, after all, with a boy's eye for fun. In the winter farmers removed the grain wagons from their wheels and set them on runners to glide efficiently over the snow. As they slid out of the elevator driveway they often had an extra passenger aboard: a young boy clinging gleefully to the back of the sleigh, catching a brief joyride before it was time to go to work again.

# CHAPTER THREE

# Toasting Marshmallows

Alex and Daisy Graham had a secret. Each night they tucked a small package beneath their pillows and each morning withdrew it before Alex went to work at the British America Elevator. It was worrisome, especially for Daisy, who had recently emigrated from civilized England to the backwoods of Birch Hills, Saskatchewan. Birch Hills didn't have a bank, so in order to keep the elevator's money safe, they literally slept on it.

Alex was an adventurer. Several years earlier he'd set out from Almonte, Ontario, following the Canadian Pacific Railroad as it snaked across Canada. In 1906 he worked on a crew in Birch Hills, twenty miles east of Prince Albert, raising an elevator for the newly established British North America Elevator Company. When the work was done, he took the position of grain buyer.

"The job was very important to Dad," recalls his oldest daughter, Margaret. "He worked hard at it."

Margaret liked to help when she could. As a small child she enjoyed taking his dinner to him at harvest time, when "long lines of horse-drawn wagons" prevented him from coming home.

"One year when the elevator was filled with grain, it burst," she says.

It's hard to imagine an elevator bursting. Cribbed construction, with planks stacked horizontally and nailed together every six inches, was designed to prevent such a disaster. To further strengthen the elevator the planks were interlocked at the corners. Two-by-eights were used at the bottom where the pressure of thousands of bushels of compressed grain was greatest. They graduated to two-by-sixes and finally two-by-fours at the top where the least pressure was exerted. When an elevator burst, it was often because its seams destabilized when corner braces in the bins broke from grain repeatedly surging against them. The corners eventually split and the walls bulged, resulting in anything from a not-so-simple grain leak to an entire section of the building collapsing. The potential problem was never resolved. In the 1980s, my husband Dale's bulk fertilizer shed in Provost, Alberta, suddenly buckled, spilling masses of loose fertilizer over the road. It was an old elevator being used for storage. The front bins were full, the back empty, and the old structure couldn't endure the uneven weight distribution. By that time a new Agro facility was under construction outside the town and plans for a new cement terminal in the same location were underway, so the old elevator stood empty and crippled until the others too were abandoned to the wrecker.

The circumstances at Birch Hills were far different. There was no question of abandoning the point. "The large amount of grain on the ground was soon cleared and a new elevator built," says Margaret.

The elevator didn't have to burst for Margaret's younger brother Jack to capitalize on spilled grain. By the 1930s the depression forced everyone—even children—to look with a creative eye for financial opportunities. When Alex loaded boxcars, the last dribbles of grain from the loading spout formed small piles beside the track. Jack and

Delivering grain by horse and wagon. Battrum, SK
PHOTOGRAPHER: TOM LEVORSON / CONTRIBUTOR: VAL HVIDSTON

his friend didn't mind putting in a bit of work to gather it up.

"They sold the pickings to a neighbour who kept chickens," says Margaret.

In 1937, with the depression at its worst, every grain handler's fear became reality for Alex. Margaret remembers it well.

"In the middle of the night the fire alarm wakened us. I looked out and said, 'It's Dad's elevator!' It was."

The town was helpless in the face of the conflagration. Devastated, they watched the flames destroy one of their most important sources of economic survival.

But those boys! Young entrepreneurs all.

"The next day was Saturday," says Margaret. A group of young boys pounded the nails out of the boards that they could handle. They sold the nails to the hardware store merchant. Then they bought marshmallows and toasted them on the elevator coals!"

Once again the British America Elevator Company wasted no time in replacing the building. The new elevator was a marvel of modern technology.

"Instead of the oil-burning engine," remembers Margaret, "electricity was used. Much cleaner and less oil smell."

Alex maintained his position at Birch Hills for forty-eight years. His family grew, literally, with the town. Margaret and her three siblings kept track of that progress in an unusual manner.

"Before bathroom scales were known," she says, "we would go to the elevator to be weighed."

One by one they stepped onto the scale and watched while Alex deftly tapped the weight across the balance beam. They stood as still as their youthful energy allowed, but a little fidgeting wouldn't have mattered.

"A few pounds here or there made no difference," smiles Margaret.

# CHAPTER FOUR

# A Load of Bull

Harold Andrews still has the silver cup that his father won in 1922 at the Stavely, Alberta, fair. Mr. Andrews's purebred shorthorn cattle were the pride of the region. That year, he mounted his saddle horse and drove a family of them from his creek valley farm near Parkland to the fair. The sire, dame, two-year-old offspring, and new calf so impressed the judges that they easily claimed first place.

Some might think it an ambitious undertaking to drive four animals down a dusty road to the fair. In truth, Mr. Andrews chose the easy route. The year before, he'd decided to show only the bull. He took it in by horse and wagon, and had a much tougher time of it.

First of all, how do you coax a bull into a wagon? The task was achieved when Mr. Andrews backed the wagon up to the valley hill so that the animal could walk straight onto it. Getting it out of the wagon in Stavely was another story, but Harold's father was a resourceful chap. He steered the team towards the elevator. After consulting with the agent, he drove his wagon into the drive shed and onto the hoist. With the wheels securely locked in place, the agent cranked the hoist. The wagon slowly lifted and tipped until

A more modern elevator than that of Mr. Andrews's time. Note,
however, that the tracks are now gone. Stavely, AB
PHOTOGRAPHER/CONTRIBUTOR: CHRIS STACKHOUSE

the confused bull had no choice but to back out onto the driveway.
From there it was a simple matter of leading it on foot to the fair. The
process in reverse at the end of the day returned the bull safely to his
pasture. With all fours planted firmly back on solid ground, he nosed
about in the prairie wool, no doubt ruminating on the strange behaviour of humans!

# CHAPTER FIVE

# Injury and Death

There was nothing different about Gordon Boettger's walk home from the Didsbury School that day. As usual, the nine-year-old was looking forward to checking the progress on the new Alberta Wheat Pool elevator going up across the tracks from his family's house. He knew some of the local men who were building it. He admired their courage, climbing high above the ground each day to sit or stand on scaffolding while they raised the elevator with lumber, hammer, and nails. Gordon watched it grow steadily. Now the day had come to install the large leg pulley that would sit in the cupola at the top of the elevator.

Gordon was too late to see the heavy pulley hoisted by rope to the top of the elevator where two men waited to roll it into position. They were guided by planks, carefully laid across the open space they needed to traverse. He was too late to see them cautiously start out. And he didn't hear the sickening splinter of wood as the boards gave way beneath them. He came upon the aftermath.

"All went down together," he says. "So hard that it cracked a large timber in the overhead bin in which they fell."

The shocked crew on the ground sprang into action, but the task before them was both appalling and overwhelming. The two men, severely injured, lay in the bottom of a bin from which there was no way out other than over the top.

"I came home from school to find them lifting these two men with a rope over a pulley that they were using to take material up for building," Gordon recalls. "They had to take the men up and then let them down the outside. It took a long time before they got them to the hospital. One man was severely crushed and died that night. The other one did survive, evidently because he dropped on top of everything."

The potential for accidents in grain elevators was great and, as already illustrated, not limited to construction crews. Lloyd Henders recalls a similar incident in a grain elevator in Manitoba.

"A load of grain had been hauled into the elevator and was waiting to be dumped into the pit. Apparently, the agent decided he should check something before starting the elevator leg. He took the manlift up to the distribution area at the top of the bin. When he failed to return, the waiting farmer began to search for him. Eventually the agent was found. He was lying at the bottom of one of the empty bins. Like most bins of that design, it had wood-cribbed walls with a few steel rods spaced to prevent the walls from bursting under pressure. The floor sloped toward the clean-out hole on one side. A wooden ladder had been left lying across a couple of the lower rods. Possibly it had been used as a temporary platform for some kind of maintenance. The man fell approximately thirty-five feet, but hit this ladder on the way down, breaking it in two. He was taken to hospital and kept for a few days in order to start recovery from a sprained ankle and several severe bumps and bruises. He had no internal injuries or fractures."

News spread quickly of this "remarkable brush with death." It was the subject of much discussion in elevator offices throughout the region. There was relief and wonder that the man was alive at all. Lloyd remembers with amusement the awkward situation in which one such discussion put an elevator agent.

"A very devout farmer solemnly declared, 'It's obvious the Lord still has some important work for Bert to do.' Another, less devout, farmer blurted out, 'That may be so, but it's obvious He has nothing urgent in mind in the next couple of weeks.'"

"Pity the poor agent," says Lloyd with a smile. "Does he laugh at one customer's offbeat attempt at humour and risk insulting his very devout customer? His valiant attempt at an expressionless demeanor was something to see."

There was no humour, however, the day the doctor said eight-year-old Arlen Capon's father might die. It was a bitterly cold January day in Storthoaks, Saskatchewan, in 1962. As Arlen passed his father's United Grain Growers elevator on the way home from school, he heard desperate, pain-filled cries for help. He ran inside and came upon a horrifying scene.

Arlen's sister, Sharon (Torz), who was then twelve years old, picks up the story.

"My father had been attempting to transfer grain from the annex to the elevator," she says. "He had gone to the extreme end of the annex, where the auger was located in a small room. He got it going and somehow his pant leg got caught and drew his leg into the machinery, chewing it up as it went along. He was dressed in multiple layers of clothes over which he wore canvas coveralls. Fortunately the heavy clothes managed to clog the auger and stop it."

The intense pain blotted out consciousness, but Mr. Capon soon revived and started calling out with all the strength he could muster. He told his terrified son to go to the fuel depot next door for help, and from there to go straight home. The men from the depot rushed over. They somehow managed to disentangle his leg and carry him home.

"When I arrived," says Sharon, "they were giving him whiskey in preparation to take him to the hospital in Redvers, sixteen miles away. The doctor said that the extreme cold kept him from bleeding to death during the time he was caught in the auger. But Dad went into shock after he was in the hospital and the doctor warned Mom that he might die. We were all grateful that he survived, but the massive scars on his legs were a constant reminder of that ordeal."

Sharon and Arlen's father escaped death that day, but not all who met danger in the elevators were so lucky. Lloyd Henders remembers the advent, in the 1950s, of grain trucks fitted with their own hoists.

"One day as one of these trucks was being unloaded in the elevator, the agent decided to check the level of the bin that he was filling. He rode the manlift to the upper level. When he came back down to the driveway scale a few minutes later he saw that the truck box had been lowered. He also saw the body of the farmer, caught in a death grip between the truck frame and the sill of the grain box. With no witnesses, the inquest could only speculate on the series of events that led to this tragedy. The farm widow and her teenaged children suffered a terrible loss."

Dave Ostryzniuk of Poplarfield, Manitoba, knows first-hand the tragedy of such a loss. He also knows how compassionate elevator agents can be.

"I was fifteen years old when my father died," he says. "My mom, sister, and I were left to carry on farming. We were badly in debt and my mom had a very difficult time trying to make ends meet. Our elevator agent was a very gentle man and did his best to try and take care of us. He always had enough room for us to deliver some grain. Weeks and months would go by and no one would be hauling grain to the elevator, because there was no room. But Mom would call the agent and he would tell her, 'Bring it in.' The community could never understand how we could always haul grain to the elevator, but we did. To this day I have a soft spot in my heart for [that] agent. If it wasn't for him, we wouldn't have survived."

There were no doubt other family, friends, and neighbours who helped the Ostryzniuks, but the elevator agent was Dave's hero. Today there are employee assistance programs in place to help elevator agents deal with traumatic events that occur in their elevators. Lloyd Henders points out, however, that fifty years ago there was no such thing. Agents simply had to deal as best they could with injury and death.

Gordon Boettger never forgot the horrific scene that he came upon on his way home from school in Didsbury that day. It didn't, however, make him afraid of heights. In fact, he preferred a bird's eye view.

CPR section foreman's house with town of Didsbury in the background. Taken by a young Gordon Boettger on one of his rooftop excursions. Didsbury, AB, 1933
PHOTOGRAPHER/CONTRIBUTOR: GORDON BOETTGER

"I was fortunate to be allowed to crawl up to the roof of that elevator to take pictures with the old Brownie box camera." (Note the men riding the rails atop the boxcars.) Didsbury, AB, 1930
PHOTOGRAPHER/CONTRIBUTOR: GORDON BOETTGER

"I was fortunate to be allowed to crawl up to and onto the roof of that elevator and one or two others to take pictures with the old Brownie box camera," he says. "One day when my chum and I were on one taking pictures, the town cop spied us. 'You better stay up there,' he hollered. 'That's the highest you'll ever get!'"

At ninety years of age, Gordon hopes that's not true.

Despite the tragic event marring its beginning, the elevator in Didsbury stood for almost eighty years. It was finally torn down in 2001, the last of Didsbury's elevators to go.

# CHAPTER SIX

# Treasures

The prospect of a visit to V.S. Foster's country store at Bremner, Alberta would light up any child's face. Even in the darkest, most destitute days of the depression, Mr. Foster could be counted on to slip a few candies into shy but eager hands. He was also the paymaster for the Bremner elevator.

"The farmers brought their grain tickets to him for cashing," says William Purdy, who worked for Foster in 1934. "They paid something on their account, then got a supply of groceries. The groceries usually totalled more than what was paid on the account, and if the farmer had children, Mr. Foster included a few candies."

One of William's duties was picking up the mail as it came through every morning on the train. The elevator agent, Tom Hanlon, was a friend of his. As William walked over to meet the mail train, he often noticed Tom coopering boxcars or moving them down the track with his "special kind of crowbar."

"You could tell when he was loading a boxcar by the chug of the big one-cylinder engine," says William.

Seeing the elevator and hearing that engine reminded William

One-cylinder engine in the basement of an elevator office.
The steps to the left lead up to the office.
PHOTOGRAPHER/CONTRIBUTOR: CHRIS STACKHOUSE

of the winter he worked for Mr. Boje, a farmer in the nearby
Graminia district, thirteen miles south of Spruce Grove.

"It was my job to load the grain box on the sleigh. It held about
sixty bushels of wheat. It was very cold that winter. After the load
was on, a small stick was slid under the runners to help keep them
from freezing down."

The same procedure was followed with the larger grain tank,
which held 125 to 150 bushels.

"It took four horses to haul this one," says William. "The next
morning we'd be off to the elevator in Spruce Grove, Mr. Boje driving
the four-horse team and I the smaller-load, two-horse team. We
walked a lot on the way to keep warm. There were green feed bundles
on the load to feed the horses when we got there, around dinnertime."

After feeding the horses, William and his boss went to the café for a twenty-five-cent lunch. Then it was over to the elevator to sell their grain.

"On the way home, Mr. Boje tied the extra team behind his sleigh," recalls William. "Back at the farm we loaded up again for the next day."

As William passed the elevators on his short walk to the Bremner train station, he was glad he didn't have to worry about those long, cold treks anymore.

Like many elevator agents of the time, Tom Hanlon and his wife Ethel lived in the basement of the office. Their surroundings were humble, but it was common knowledge that elevator agents were among the few who could count on a steady paycheque. In 1930 they made between $100 and $150 dollars a month. As the depression deepened, wages were reduced, but many still made between $70 and $100, and it came regularly. In addition, because the elevator operated as a business, there was always the possibility of money on the premises. Whether it was one of the locals or simply a boxcar hobo passing through, someone clearly felt that good fortune should be shared.

"Once, while Tom and Ethel were out at some social function," says William, "somebody broke into their place hoping to find some cash. It wasn't kept there, but they stole some of Ethel Hanlon's treasures."

Were the treasures sold for cash? Or were they sent home as small recompense to a grass widow who'd lost her man to the dust and the rails in an endless search for work? No one will ever know.

Today, says William, "Bremner is no longer on the map. It was at the crossroads of Highways 16 and 21. V.S. Foster sold the store building to a local farmer and moved to Vancouver. There is an overpass at that spot now."

# CHAPTER SEVEN

# One Bushel at a Time

Evelyn Battle (now Dickie) was just coming into her teens when the Great Depression descended with full force on the Delia district in central Alberta. She remembers the "grasshoppers, dry windy weather, and everlasting dust storms" as though they were yesterday.

"Money was hard to come by," she recalls, "but my father, by some miracle, always seemed to have a bit of grain in his bins to last through the summer."

Evelyn and her seven siblings took for granted their responsibilities pitching in on the farm.

"At least once a week it was my duty to bag a bushel of grain," she says, "and drive four miles by horse and buggy to sell my wheat to a kindly elevator agent. He always weighed it carefully and paid me as much as possible for every kernel in the bag. Those kindly managers will live in my memories forever."

After selling her bushel of grain at the elevator, Evelyn steered her horse and buggy to the Delia store where her mother had an arrangement with the storekeeper. Like many farm women, she kept chickens, selling the eggs for extra money. She also churned butter.

Delia, AB
PHOTOGRAPHER/CONTRIBUTOR: CHRIS STACKHOUSE

"My mother always sent along four or five pounds of real farm butter, a dozen or so eggs, and milk and whipping cream. This was all deducted from the grocery bill."

The industrious Battle children plunged into the spirit of making ends meet.

"In the berry season we picked gallons of saskatoons," says Evelyn, "which we sold to the town people."

Life was good, as far as she was concerned. She describes the Dirty Thirties as "a part of my life that holds fond memories. And the grain elevators were certainly part of them. We were so proud of our eight beautiful elevators standing tall for all to see for miles around."

But as the years went by, those Delia elevators disappeared one by one until, in 2001, the last one burned to the ground.

"Wooden grain elevators are dear to my heart," says Evelyn. "I feel so saddened now when I pass a prairie town that has had to see part of their history fall by the wayside."

# CHAPTER EIGHT

# Full Circle

If Earl Cruickshank had been counting, he would have lost track of the number of times he trudged with his wheelbarrow between the Alberta Wheat Pool elevator in Botha and the shed in his backyard. It didn't matter. He knew a golden opportunity when he saw one. The well-worn driveway in his elevator had echoed with the clatter of thousands of hooves. Now it was being torn up and replaced. All those planks would go a long way towards keeping the home fires burning that winter. The depression was a hard teacher. One of its biggest lessons was that nothing must go to waste.

Elsie Cruickshank was glad of the extra fuel for her stove. Neatly cut, split, and stacked in the woodshed, the supply seemed endless. Yes, endless. Every day she stubbornly went out to retrieve more for the fire and every day she and her family tried valiantly to close their noses as the wood warmed and burned, unlocking all the aromatic secrets of its past: years and years of horse droppings and urine.

The depression was just around the corner when the Alberta Wheat Pool bought the Farmers Elevator in Botha, just east of Stettler, Alberta, and Earl moved into the community as the new agent. A few weeks later he married Elsie and she joined him in a rented house a block and a half from the Pool. They were happy in their new lives and even managed for a while to hold the pall of the depression at arm's length. But its clutch was inevitable.

"One of the few holidays Mom and Dad took during the depression years was to Gull Lake," relates their daughter Marie Barnec. "When they returned, they found the monthly paycheque in the mail but it had been reduced because of the hard times. There were no previous warnings, just fewer dollars."

Earl Cruickshank in the driveway of his elevator. The well-worn driveway had echoed with the clatter of thousands of hooves. Botha, AB
PHOTOGRAPHER: UNKNOWN /
CONTRIBUTOR: MARIE BARNEC

Earl wasn't one to let hardship get the better of him. Across the road from the elevator stood a plot of land ideal for a large garden, which he wasted no time planting. He kept an eye on the garden from his elevator and an eye on the elevator from his garden. A farmer stopping by during a slow spell invariably found Earl across the street with his hoe.

Earl also kept chickens.

"He had laying hens in the bottom of the elevator," says Marie. "The space may have at one time been used when they had the fanning mill or the grinder, for cleaning grain and grinding chop in the elevator."

It wasn't uncommon for elevator agents to keep chickens. There was, after all, plenty of spilled grain lying about, both in the loading area and in the cavity under the elevator. Eventually, however, elevator companies put a halt to the practice.

It was just as well that Earl gave up his chickens. Business at the Botha elevator was steadily increasing. In 1940 a new annex was built to accommodate the additional yield. New or not, the roof leaked. Wet grain deteriorates quickly, so when it rained, Earl had to set out buckets. And when it rained through the night, he had to get up and go down to the elevator to check them.

One night he wearily climbed the ladder on the outside of the annex and stepped through the upper door. He played his flashlight along the narrow catwalk that extended over the bins. He expected to see the buckets he had placed to catch the rain, but instead the light caught a pair of gleaming eyes. Earl froze. Clearly there wasn't room for both him and the porcupine on that catwalk, and Earl decided quickly it was best to be polite. He withdrew. The next day, from a safe distance, he and a customer stood by while the overnight guest slowly lumbered from the annex and waddled off down the road.

Because of the risks involved, elevator agents didn't like taking in grain that wasn't dry. If it didn't meet minimum moisture standards, it was considered tough or damp. This grain heated quickly in storage, bringing on mites and rusty grain beetles, causing spoilage and, if left untreated, the risk of fire. It took tremendous time and effort to cool heated grain, turning it constantly from one bin to another to keep air flowing through it. If it was really hot, a board was placed across the back hopper so that the grain was split in two, allowing even more air to circulate as it flowed out of the bin into the pit before being elevated into another bin.

Agents learned to recognize excessive moisture content simply by the feel of the grain. If their hand slid easily to the bottom of a bucket of wheat, they knew it was pretty dry. But if they had to push their hand through, chances were it was tough. They could also tell by how soft the grain was when they bit into it.

One way my husband distinguished tough from dry grain was by watching our dog Jiggs, who accompanied him to the elevator almost every day for twelve years. Jiggs loved to eat wheat. As it cascaded from the grain truck into the pit, he grazed on the kernels that spilled over onto the floor. He didn't bother with dry wheat. It was too hard and crunchy. But he couldn't get enough of the softer stuff. (I couldn't blame him. I'd learned as a child that wheat makes great chewing gum—if you have the patience to keep chewing, it forms a soft, elastic mass in your mouth.)

"Jiggs would stay and eat wheat until you swept it out from under his nose," says Dale.

When a farmer and buyer couldn't agree on the moisture content of grain, more scientific measures were needed. Art Braun, of Altona, Manitoba, says, "As long as the threshing was done from stooks, which stood up off the ground, I can't recall having to worry about tough grain. But with the advent of swathers and combines, you found the need to have moisture testers."

This was because swaths, lying on the ground, were so much more vulnerable to rain and frost. Over the years various moisture-testing devices were invented. One early method involved heating a measure of grain in a small oil-burner with a temperature gauge attached. As the grain cooked, the needle on the gauge crept upward. When the needle stopped and began to regress, the agent knew there was no moisture left in the grain. The reading on the gauge at the point that it stopped indicated the moisture content of the sample, and therefore the load. A reading of 14.5 for wheat was considered dry. Anything over that was tough.

This method was generally only used when the farmer and agent

Alberta Wheat Pool elevator. Botha, AB, 1958
PHOTOGRAPHER: UNKNOWN / CONTRIBUTOR: MARIE BARNEC

were in serious disagreement. Not only did each test take about half an hour to complete, but the smell of heated grain was ghastly.

Some farmers practised creative ways of getting around moisture testers. Marie remembers that in Botha her father Earl "had one customer whose wife kept the grain sample in the oven overnight before it was brought to the elevator to be tested for moisture.

"I'm not sure how Dad knew this was going on," she says, "but I know he mentioned more than once this yearly happening."

Of course, no agent ever got away without having to deal with extremely poor quality grain at some point. Art Braun recalls what happened in such a case.

"If you had carloads of this type of grain, you could ship it and hope it wouldn't be rejected at the terminals. Or you could try and blend it into dry grain. But how much or how little not to spoil the whole car? If you could do this without getting caught, it should be a feather in your cap in the eyes of the company."

During harvest season elevator agents were on call twenty-four hours a day, seven days a week. I remember how exciting it was. Everyone looked forward to getting the harvest in. Dale and other agents didn't mind keeping the elevator open until midnight or later, doing moisture tests and taking in grain. When they were at home they were still on call, ready to go back on a moment's notice. Many nights farmers came to our door with a sample for testing or a load to be dumped. By this time faster, cleaner moisture-testing methods had been developed. In Paradise Valley, Dale set up an extra moisture tester in our garage, saving many trips back through the path in the bushes and down the track to the elevator.

This spirit of co-operation lasted as long as there was grain in the field. When harvest was over, the elevator man returned to some semblance of the forty-hour workweek for which he was paid. Unfortunately, the odd farmer couldn't grasp that agents were eating, sleeping humans like the rest of us. Depending on his mood, Dale either shook his head, laughed, or got downright annoyed when the occasional farmer dropped his truck off at the elevator at noon, then went to the local diner for lunch, expecting Dale to unload over his own lunch hour.

Earl Cruickshank once paid a high price for being accommodating.

"Mom and Dad had guests playing bridge one evening," says his daughter Marie, "and a knock came to the door. It was one of Dad's customers, and he had a load of grain. It wasn't harvest season, when Dad was used to late-night calls, but he pulled on his coveralls and went with the farmer to the elevator."

Dumping grain at that time included climbing into the truck box with a broom to sweep out the corners. Earl had just finished and was jumping out when he lost his footing and fell heavily to the driveway floor.

"Are you okay, Earl?" asked the farmer.

Dazed from the unexpected fall, Earl replied, "I'm not sure."

He got up unsteadily and stumbled to the scale. As soon as the empty truck was weighed the farmer got in and took off, not bothering to ensure Earl's well-being, or even to offer him a ride back home.

"In those days there wasn't a phone in the main part of the elevator," says Marie. "Dad had to hobble to the office and call Mom to come and pick him up. When she arrived and saw the size of his ankle, she took him straight to Stettler, our nearest large centre. They X-rayed the ankle and discovered it was broken."

Marie is quick to point out that "not all of Dad's customers were like that one." The majority of farmers appreciated the work of the elevator agent. This was often demonstrated in grand fashion at Christmastime. Sitting in our cupboard still are bottles of spirits given to us by generous farmers. We also received chocolates, baking, and even the occasional turkey. Once, we were presented with half a pig, butchered and wrapped, ready for the freezer.

Rose Hooley, whose father, Thomas Henry Small, was the United Grain Growers buyer in Lamont, Alberta, for thirty-seven years, remembers the farmers as "sincere friends, always ready to help if anyone was in need."

Rose's family made sure it was a two-way street.

"The elevator sold chick feed, pig feed, all kinds of feed, and also animal salt and hundred-pound bags of flour," says Rose. She helped her dad, "carrying feed from boxcars to shed, coopering boxcars, unloading coal into storage sheds, and then helping the elderly farmers load coal into wagons for fifty cents, or sometimes nothing at all. The exchange was always very rewarding. A nice fat chicken for Sunday dinner or cream to go along with wild mushrooms.

"Many times Father sent my brother and me under the elevator to collect the grain that had spilled from the hopper. We were given a nickel for a two-and-a-half-gallon pail. We thought we'd hit the jackpot. Two pails bought us an ice cream float.

"My mother also played a big part in elevator life. Not only did she help Father with coopering boxcars, but also, many lunches were served in the little kitchen of our house across the road. Mother was an excellent cook and there always seemed to be enough food for that other farmer who happened to be around at lunchtime. On the return trips of many of these farmers, there was always a gift of a nice fresh roast pork or something for Mother's kitchen."

In Botha, Marie Barnec remembers "the farmer who always brought a quart of cream when he had finished hauling his grain in the fall. Other farmers brought meat or butter, sometimes even candy."

During the war, when there was a critical shortage of manpower on farms, Earl helped out by cutting and stooking grain at harvest.

"His pay one fall was meat when the farmer butchered a pig," says Marie. "Unfortunately the pig was an old boar and that was one time my mom did throw food out."

Still, the intention was appreciated.

"I suppose now it would be imprudent to accept those kinds of offerings," Marie reflects. "But in those days they were given as a simple thank you."

Earl retired in 1966, having spent thirty-nine years in Botha. It was the company record for number of consecutive years served in one point. Shorty Skocdopole was on hand for the occasion. He had been Earl's first customer in 1927. Now he would be his last. In a re-enactment of his first delivery, Shorty hauled a load of grain into Earl's elevator by team and wagon. Earl's career had come full circle.

Although there is no longer an Alberta Wheat Pool elevator in Botha, a visit to the Barnec farm shows that Marie never strayed far from her roots.

Earl's last grain delivery in Botha: an exact
re-enactment of his first. Botha, AB, 1966
PHOTOGRAPHER: UNKNOWN / CONTRIBUTOR: MARIE BARNEC

"The big hopper that was used to weigh grain before it was loaded in the freight cars found its way to our farm, and we used it for years as a chop bin," she says.

Near her garden stands an original Pool elevator sign, personally delivered from the Bashaw demolition site, on condition that the Barnecs had a front-end loader to get it off the truck.

"My husband is building me an elevator to stand beside it," she says. "The elevator will be built from salvage of the Botha elevator. Also included in the salvage are grain-worn boards from inside the bins. Those someday will be made into a bench for my office."

Even Marie's scrap paper is memorabilia.

"We always had a supply of old elevator record books to use as scrap paper, and still have some today."

# CHAPTER NINE

# Peanuts

In mid-December, in the heart of the depression, a farmer in Altona, Manitoba, rattled up the driveway of the Federal Grain elevator with a hundred bushels of oats on his triple box. Art Braun watched as his father carefully weighed the oats, dumped them, and took a sample for dockage and grading. He made out a cash ticket for as much as he could possibly pay.

The farmer looked at the ticket and his face fell. Eleven dollars—and Christmas was coming.

"He commented," Art recalls, "that because he had a large family, it would be barely enough to buy peanuts."

Art's ten-year-old imagination tried to wrap itself around the idea of having only peanuts for Christmas.

"I was never privy to what my dad's salary was," he says. "But I believe that a grain buyer's salary must have been quite reasonable. During the depression I did not feel that we were poor. The salary came in regularly, and we went for holidays to Winnipeg Beach with other relatives and friends."

Still, the man with "barely enough to buy peanuts" was better off than many farmers who were bypassing the elevator system and shipping their own grain by producer cars.

"They ended up owing the railways more than they got for their grain."

Aside from world market conditions driving grain prices to all-time lows—as little as twenty-five cents a bushel for wheat—much of the problem was drought. Not only did extreme dryness severely compromise the quality of the grain, it also reduced it to almost half its normal weight. As his father's helper, one of Art's jobs was to climb into the boxcar while it was being loaded and shovel like mad into the corners.

"The grain didn't have sufficient momentum on its own to reach the end," he says, "but you still had to get in the minimum weight."

Art also climbed into flat-bottomed bins to clean out the grain left behind when they were emptied. It was tedious, dirty, back-breaking work.

"You were sure that there were at least a thousand bushels to shovel," he says, "even though it was probably only two-fifty."

As he worked, grain dust swirled up around him, settling like fine snow on his hair, clothing, face, and hands. It invaded his eyes and crept up his nostrils. In later years elevators stockpiled dust masks for use whenever grain was handled, but during the Dirty Thirties there was no such thing.

On rainy days Art was assigned to the top of the elevator to clean up accumulated grain dust. Of course he used the man-lift.

"The counter-weight was set for the buyer's weight," he says. "So if you weighed half as much, your ride up could be quite fast and you needed to know when to brake. Going down was another matter, when you literally had to pull yourself down. But it certainly beat climbing the dusty ladder."

Art wasn't the only family member who helped in the elevator. In

Pitching corn from a horse-drawn wagon into bins created
from snow fencing. Altona, MB, 1941
PHOTOGRAPHER: UNKNOWN / CONTRIBUTOR: ART BRAUN

fact, the family had suffered a terrible tragedy several years earlier.

"I had an older brother who at the age of thirteen was killed at
the elevator. At that time the engine was still in the engine
room/office and the driveshaft led from there into the elevator. My
brother was going to check the fuel tank, which was on the track
side. As he ducked to go under the driveshaft, which was uncovered,
his clothing caught, and before anyone could disengage the clutch,
he was hurled to the ground and died. In those days there were no
big settlements, no big lawsuits. I gather that the company paid for
the funeral expenses and life went on."

So it had to. There were crops to grow and harvests to bring in.
During the war, farmers in the Altona area began growing corn.

"The corn ears were hand-picked into a wagon and brought to
the elevator," says Art. "Another brother and I built sixteen-by-six-
teen foot platforms and with a roll of snow fencing created a bin.
When the first layer was full, we put up another tier. The farmers
were required to fork the cobs from the wagon into the bin. Toward
the end of January a local farmer owning a corn thresher came in to

thresh all the corn we bought. This method was used for only a few years and then regular combines were used to harvest the crop."

It wasn't always men who hauled to the elevator by horse and wagon.

"In those times," says Art, "quite young children would bring the load of grain to the elevator. I remember having to go down the driveway and lead the team in."

In 1943 Art joined the air force. By the time he returned he was a man, and the whole world had changed. He began working for Federal Grain in 1945, the same year his father retired from the company.

"There was a time when shipping seemed to be a problem," he says, "and elevator space was augmented by the big balloon annex. A single bin held about twenty thousand bushels. Care had to be taken that the grain that went in was clean and dry, because there was no easy way to empty the bin. In spite of that, mites became a problem and the bin had to be checked with a long probe."

It was one of Art's first jobs with the company to travel to Federal elevators within the region, treating grain stored in balloon annexes with a special gas designed to "combat the infestation."

"I was a buyer for only three years," he adds. "I felt that there were better ways of earning a living than to go from working sixteen- and eighteen-hour days during the busy season to the utter boredom of the off-season."

Art also experienced many frustrations on the job—frustrations echoed by grain buyers everywhere.

"In the off-season, you might go days without any work. Of course, if you did leave the office for a short time, you'd come back to have someone say they'd been waiting for at least an hour. The farmer usually thought too that the buyer was out to cheat him, either by dockage or weight. I remember one customer who wouldn't deliver grain to the elevator from the threshing machine. He stored it at home, then later bagged and weighed each load as he brought it in. Was that a pain to unload! I changed my occupation to being the manager of our local credit union and spent forty-two years there until I retired."

Nevertheless, Art mourns the passing of the wooden grain elevators.

"For the most part our town had three: Lake of the Woods, Ogilvies, and Federal, with the Pool coming in later. We lost our last elevator a few years ago and the nearest one now is fifteen miles away. In a fourteen-mile stretch (three towns), we have seen twelve elevators disappear in the last several years. Granted, the mode of delivery having changed from horses with a sixty-bushel load to the present large semis, it still seems sad to have lost these sentinels of the prairies."

# CHAPTER TEN

# Rough Rides

Teena Feniak states it simply: "I am a farmer's daughter and a farmer's wife." Nevertheless, until she was a teenager, she'd never set eyes on an elevator. She lived in Worsley, Alberta. It was forty rough, bumpy miles from Hines Creek: the nearest town, the nearest elevator, and the end of the railroad line. In the late 1940s and early 1950s it was a formidable distance and her family, like many others, didn't own a truck.

"In those days," she says, "people helped each other with things such as hauling grain."

Teena will never forget the red-letter day that she saw the town for the first time. She needed school books. Her uncle was going to Hines Creek in his Diamond T truck and she could go with him.

"He was taking a load of grain for a neighbour," she says, "who was also along."

Very early on a mid-week morning, the three set off in the big truck. Teena sat in the middle, and even though the ride was rough over unconditioned roads, she watched with wonder and anticipation as the world bounced by. Her eyes widened as they approached the elevators. She'd never seen anything so huge!

Teena got the books she needed, and the trio started back in good time. But the novelty of the trip had worn off a little.

"This time, guess who was *drafted* to sit near the *drafty* door in that truck?" she moans. "After forty more miles I was deposited, cold, somewhat late, and shook up, at the school for the rest of the day."

While many farmers still hauled by team and wagon, mechanized methods of transportation were becoming more popular. There are no people more innovative than farmers, and before the large grain trucks we are familiar with came along, there were many homemade substitutes. Claude Stevens describes one that he and his father put together during the Second World War.

"We got an old 1929 Model A Ford truck for two hundred dollars. We attached it to an eighty-bushel grain-tank-style box and two transmissions, one behind the other. With both in low, it pulled like a tractor."

What formerly took Claude two hours by horse and wagon now took only half an hour.

Lloyd Henders adds that "during the 1950s a farmer would buy a used cab and chassis unit and mount a grain box that held approximately two hundred bushels. A frame-mounted hydraulic hoist was installed to dump the load. That was more efficient than using the elevator air hoist on the front axle of such a long vehicle."

But many trucks didn't yet have their own hoists, so the elevator air hoist was used, lifting the entire front of the truck, including the cab, and tipping it back to dump it just as though it were a wagon. My husband, Dale, was a small boy in the 1950s. He never missed the chance to accompany his dad to the Federal Grain elevator in Herschel, Saskatchewan. He was particularly thrilled to be allowed to sit alone in the cab while the truck was hoisted high into

The air hoist lifted the front of the truck, tipping it
back to dump it. Salvador, SK
PHOTOGRAPHER: UNKNOWN / CONTRIBUTOR: WALTER ZUNTI

the air to dump its load. He loved the sensation of rising up past
the men in the driveway, at the same time tipping steeply to point
towards the sky. He was in his own little world while the truck was
dumped. The ride down was sensational. Slow at first, then almost
stopping before a sudden drop flipped his stomach into his throat.
After that a soft landing eased him back to earth. He says it was like
being on a fair ride.

Teena Feniak's small niece had a different reaction to the same
experience. Like her aunt, she was not very familiar with elevators
when she kept her dad company delivering grain to town one day.

"She was alone in the truck when the front end was going up,"
says Teena. The terrified girl screamed and screamed. "Finally, they
reversed the process and let her out of the truck. She wasn't aware,

as were the guys standing around, that what was happening was quite routine in a grain elevator."

Truck-mounted hoists eventually became the norm, but not until the early 1970s were they considered standard enough for air hoists to be removed from the elevators. Perhaps that was still a little too early.

"I remember that time," says Leonard Wold of Alliance, Alberta. "Sometimes the first truck hoists failed and the load had to be shovelled off by hand."

This happened to him on one occasion.

"I also remember twice when farmers forgot to lower their boxes before driving out of the plant. There was damage to the elevator and truck box once, but in the other case there was loud enough yelling from the agent and bystanders to alert the farmer."

Perhaps one such bystander in one such elevator was Teena Feniak, for after she married, she became very familiar indeed with the towering structures.

"I often accompanied my husband on his trips from our granaries to the Whitelaw elevator," she says, "where he unloaded sometimes several truckloads of grain a day. I saw grain elevators every day in that small town. In fact, we lived near them. Of course, we were thus privileged to hear (and feel) the train rumbling through as well. But it wasn't always that way."

CHAPTER ELEVEN

# Three Ladies

The transition from horsepower to mechanization on the farms took place slowly over many years. During the Second World War, the transition from manpower to womanpower happened much more quickly. Mabel Wheeler (now Hobbs) took it fully in stride. Not only was she used to farm work, she enjoyed it.

"In our family there were three girls and then a boy," she says. "I was the middle girl. The older one helped my mother, but I did a lot of chores with my dad. I was probably nine or ten years old when I accompanied him to the elevator for the first time. Our farm was on the bench of the Cypress Hills, eleven miles south of Carmichael, Saskatchewan. It was a long trip with a team. I remember looking down on the flat and seeing the strips of summerfallow and stubble fields looking like a giant patchwork quilt."

By the time the war began in 1939, Mabel was a teenager and her father made the decision to move the family closer to Carmichael so that his children could attend high school.

"We moved to a rented farm one half mile south of the village," says Mabel. "I admired the view of the five elevators standing dark

Threshing season. Mabel shovels grain in the back of the truck,
taking it straight from the separator to the Pool elevator.
Note the water jug sitting in the shade of a stook.
Carmichael, SK, 1943 or 1944
PHOTOGRAPHER: UNKNOWN / CONTRIBUTOR: MABEL HOBBS

against the sunset sky many times in late autumn as I drove the
cows home for evening milking."

Tending the cows was only one of Mabel's many chores.

"The war was on, and men were scarce," she says. "When the
men weren't there to thresh, the teenagers took over and did it. Our
threshing crew was made up of young boys—fifteen, sixteen, or seventeen years old."

Mabel was right in there with them.

"We all worked the long hours that threshing demanded. I used
to shovel grain back from the threshing machine and shovelled it off
into the bins with the old grain scoops."

She also shovelled fifty bushels at a time into the wagon box,
taking it straight "from the separator to the Pool elevator."

When Mabel was finished with her own work, she helped a
neighbour.

"His farm was about five miles south of town," she recalls. "He had a grain tank which held about ninety bushels of grain. I would make two trips a day from his farm."

There were no such things as holidays during those years. Her father didn't even own a car until after the war.

"After he bought the car, we used to go to Gull Lake for the theatre on Saturday nights. But before that we didn't go. If we did, it was with team and buggy for the May 24 Victoria Day holiday, and that was about it."

It was just as well to stick with horsepower during the war years, when both gas and tires were rationed.

"In 1944–45 our agent, George Beach," says Mabel, "boasted that he had three ladies hauling grain to his elevator. One was a grandmother helping her son. She drove a small truck. The second lady was a widow who was trying to scrape a living from the farm after the death of her husband. The third was me, a young teenager, who was the only one who drove a team of horses."

The last of those five elevators—Pioneer, Federal, McCabe, Western and the Saskatchewan Wheat Pool—disappeared fifteen years ago.

# CHAPTER TWELVE

# Lead in Her Pants

"When I drive through the country," says Verla Nevay of Rosetown, Saskatchewan, "and happen to see an old elevator standing alone, like a deserted friend, I close my eyes and once again visualize the big engine popping away, the long goosewalk between the elevator and the office, the belts purring as they run the interior parts and the leg carrying grain from the pit up to the storage bins."

It doesn't take much for the memories to surface. When Verla married Len McLellan in 1948, their first home was in an elevator. They'd planned to leave Saskatchewan to seek their fortunes in Ontario, but when Len was offered the job of grain buyer for the Canadian Consolidated Company in Valley Centre, he decided he had nothing to lose by trying it.

"Some companies had separate houses for the operators away from the elevator grounds," says Verla. "Our living quarters were built into the office and consisted of a kitchen and a bedroom."

Before Verla and Len could move in, they had a nasty job ahead of them.

"The company supplied all the items needed to clean the place," she says. But they didn't supply the elbow grease. That was left to Len, Verla, and Verla's sister Lois.

"This place had been empty for a year and a half," she says, "so we faced many distasteful surprises. Dirty, dusty, smelly, dead mice, and a half a case of beer, which had frozen, and the smell permeated the place. I kept thinking that this was not really what I expected my first home to be like."

The three persevered. They scrubbed everything in sight, following up with a paintbrush, which they even applied to the floor. They repaired the windows and Verla hung curtains.

"Each day made it look and smell better," she says.

Next was locating the necessities for a home.

"We got an old kitchen table and two chairs from my uncle. They were brought from North Dakota when my grandfather migrated to Saskatchewan. We got a foldaway cot to sit on and use as an extra bed in the kitchen/living room. We bought a kitchen range and a complete bed unit from a person in Herschel, and along the way we scrounged other knick-knacks to set up housekeeping."

Gradually the two shabby rooms transformed into a tiny home that Verla and Len could settle into.

Only one thing troubled Verla: she shared her living quarters with the engine that powered the elevator.

"That big thumper ran most of the day and it kept my pots and pans on the stove rocking and rolling."

Not only was it noisy, but Verla didn't trust it. Worse, she was afraid of it!

"If that engine went on the fritz, nothing else would work. Len would prime the motor, step up on that big wheel, and give a pull to turn it so it would start. When I first saw that huge wheel I was petrified. My uncle was killed when he became entangled in one. I thought about this every time I saw Len go to start the engine, but he managed to do the right thing. No accidents there."

Verla was trained as a schoolteacher, so to supplement Len's income of eighty-five dollars a month, she occasionally did some substituting. For the most part, however, her help was needed at home.

"The wives of the grain buyers were ready to adapt to a new way of getting along," she says. "We did the book work, fire prevention reports, and cleanup jobs. Some of the wives could weigh out a load if the operator was away, but [they couldn't write a] cash ticket. That was a no-no."

Verla became accustomed to helping Len. In 1950 he switched over to the United Grain Growers elevator, and while Verla was in the hospital having a baby, helping hands moved them lock, stock, and barrel into the UGG company house in Valley Centre. Soon she was back on the job as Len's main assistant, willing to help out wherever she was needed.

One Saturday she got a frantic call from the elevator.

"Len needed me to help him move a bin of hot wheat. I got there and climbed into this hot bin to chisel, chop, and shovel the wheat into the spout where it was carried down the leg and moved here and there to cool off. This bin of grain was really hot. I could feel the heat through my shoes. I didn't stand in one place too long. We got the grain cooled and moved about, but it all had to be done again in the morning to keep the grain aired out and cool."

Suddenly their plans to take in a ball game in Rosetown that afternoon weren't so exciting.

"We did go to town for a while, but we were tired out and worried, as these hot bins can cause fires."

Verla also coopered her fair share of boxcars. Often the train spotted four on Friday, expecting them to be loaded and ready for pickup on Saturday. This imposed a tremendous time crunch on Len, but Verla was always ready to help. It didn't trouble her in the least to climb into a dank car and box herself into it with grain doors. What did bother her was climbing out.

"There was no way I could boost myself up and out of that car," she recalls. "So there was always a ladder by the car I was in. It was a standing joke: 'Verla needs help; she has lead in her pants!'"

Verla laughed with the rest of them, but secretly the wheels were turning.

"I often think about that manlift," she says. "So many times the grain buyer stepped on that lift and went up to check a bin he was loading into. My sister was light and skinny, so she could ride up with Len on the lift, but not Verla! She still had lead in her pants. I was determined to get up to that third level on my own."

Verla took a good hard look at the ladders leading to the top of the elevator. She grit her teeth and put her foot on the first rung.

"That was the longest and highest trip I ever attempted," she states. "I made it, but never again. I was a nervous wreck when I got back to ground level."

The ladder descending the dark shaft below the trap door in the driveway was almost as frightening. Nevertheless, Verla regrets the day she wasn't there to use it. It led to the dreaded boot, a narrow space twenty or more feet underground in the dusky, creature-inhabited bowels of the elevator. Here the endless belt that comprised the leg reversed its journey, inverting its cups to scoop grain from the pit as it ascended back to the top of the elevator. Access to the boot was necessary for several reasons. One was to unplug the leg if it became clogged with grain and stopped moving. This is what happened one day when Verla was substitute teaching and Len was alone in the elevator. He opened the trap door and dropped a trouble light down the shaft so that he could see his way. Then he stepped onto the dusty ladder and into the depths. Once below, he opened a hatch exposing the leg and thrust his hand in to clear the grain that was plugging it.

"As he moved the grain, the leg started moving down," says Verla. "The cups on the belt were full. One came down on his hand and pinned him."

With his arm thrust awkwardly through the hatch and his hand smashed heavily, painfully to the floor, Len couldn't move. Way up above him he saw the patch of light exposed by the open trap door, but he knew there was nobody up there to help. All he could do was wait and pray that somebody would happen to find him.

"There were three other grain buyers in Valley Centre at this time," says Verla, "and these guys helped each other in many tasks."

To be sure, they were competition, but a bond of camaraderie existed along elevator row. Many a grain buyer can also attest to an eerie psychic connection.

"After so long the guy next door in the Searle elevator sensed something was wrong. He came and found Len in this predicament. He got others to come and help get Len untangled and up on ground level again. His hand was badly crushed, but there were no bones broken. You can bet Len gave a prayer of thanks for his life that day."

The couple also thanked God for friends.

"I've said many times," repeats Verla, "friends are priceless!"

And friends they had. By this time their family included two small sons, and they were heavily involved in community events.

"Len always enjoyed sports," says Verla. "Softball, hockey, curling, skating—and he was a really good dancer. We got involved with each event as the seasons rolled. The men had a hockey team. They flooded and made a skating rink for everyone to enjoy. There were ball games in the summer, and each small town had a team."

Verla and Len had a good life with their growing family. Verla was in heaven when UGG decided to build a new addition onto their company house.

"Construction completed, we had a lovely roomy kitchen and a

Residence consisting of a kitchen and a bedroom built into the office of the Canadian Consolidated elevator.
Valley Centre, SK, 1948–49
PHOTOGRAPHER: UNKNOWN /
CONTRIBUTOR: VERLA NEVAY

United Grain Growers company house. Verla was in heaven when UGG decided to build a new addition.
Valley Centre, SK, 1950
PHOTOGRAPHER: UNKNOWN /
CONTRIBUTOR: VERLA NEVAY

new bedroom. We were given the job of doing the interior finishing, so we filled cracks, sanded, and painted. Oh, it was a real palace! We had a basement dug and a new furnace for heat."

What a long way from the two little rooms in the office of the Canadian Consolidated elevator.

But it wasn't to last.

"That year," says Verla, "Len became very ill with a kidney infection."

It was so severe that he was laid up in the Rosetown hospital for three months. It was a trying time, but United Grain Growers held his job for him. The family was overjoyed when Len came home again.

Soon afterward he received word of a transfer to Rutland, ninety miles away.

"I was happy for him," remembers Verla, "but I didn't feel too good about leaving our nice new home. We went to Rutland to look the place over, and after June exams we packed up and moved. We had a cat in a box and the canary in a cage in the back seat of the car. Everything arrived safe and happy."

"Rutland was a small town, but the house was well-kept, and we found the folks very friendly. It had conveniences too. Grocery store, post office, mail service every day, church services in the community hall, parties, and dances. The school was right beside us in the next yard. We were right at home in no time."

Verla fondly remembers the social life.

"The house was well-kept. We were right at home in no time."
Valley Centre, SK, 1959
PHOTOGRAPHER: UNKNOWN /
CONTRIBUTOR: VERLA NEVAY

"Dances were our special outings. At Rutland we had Ackerman's Orchestra and we took in all the dances we could. We went every week to Senlac, Winter, Unity, or Macklin. We danced miles and miles."

The Ackermans once proved their worth for more than just music. It was a sleepy Saturday in Rutland, quiet because most of the townspeople were twenty miles away in Unity doing their weekly shopping. Afterwards they'd go for dinner or a show. Len was at work, but there wasn't much happening. He was looking forward to the dance in Senlac that evening.

"At about 2 PM a monstrous big truck moaned into the elevator," says Verla, "with a big flat deck loaded with bagged fertilizer."

Len took one look and knew right away that they'd never make it to the dance. It would take ages to unload that truck. He called Verla to break the bad news.

"I took the boys, and we walked to the elevator," she says. "Len was carrying bags of fertilizer to the storage bin."

As usual Verla rolled up her sleeves and pitched right in. The

UGG elevators, Rutland, SK, 1959
PHOTOGRAPHER: UNKNOWN / CONTRIBUTOR: VERLA NEVAY

two of them plugged away while the small boys played.

Then Tony Ackerman came by with his two boys.

"When they found out our plight," says Verla, "nothing would do but they'd help." With triple the manpower, spirits soared, and with renewed energy they all set to work.

"We carried bags for about two and a half hours. When it was done and the Ackermans were leaving, they waved and called 'see you guys tonight.' We went to the house to clean up. We had supper, a bath, and a wee rest. Then we bundled up our boys and went dancing. By nine o'clock we were doing our favourite polka."

Verla has special memories of those dances. Local music, midnight lunches, perhaps a little homebrew out back for the men, and dancing far into the night.

Saskatchewan Wheat Pool, Rutland, SK
PHOTOGRAPHER/CONTRIBUTOR: CHRIS STACKHOUSE

"When the boys got tired, their dad brought in a blanket from the car and put it behind the piano up on the stage. They'd be sound asleep in no time."

Verla and Len were happy in Rutland, but their lives were destined to change once more. In 1961 Verla's father passed away.

"Len gave up the elevator to come home and operate my dad's farm. In 1962 we added to our family with a lovely baby girl. Len passed away in 1974. Now Valley Centre is wiped off the map. There's nothing left to tell of the happy times and good friends that once made the town. It's a cow pasture now."

# CHAPTER THIRTEEN

# A Great Collection of Empty Bottles

If Gerald Hallick had known that being an elevator agent would put him in the hospital twice, he might never have taken the job. Even his introduction to the career was somewhat violent.

After finishing high school, he began as a helper in the Manitoba Wheat Pool elevator at Starbuck, his hometown.

"It was not my intention to make it a permanent job," he relates, but his cousin's husband needed help and Gerald was capable and willing.

One day after work, as the two men unwound in the local pool hall, a farmer approached.

"He'd been drinking," says Gerald, "so he was brave enough to argue with my boss about the grade he got on his grain."

Tanked-up characters with loose lips aren't the most delightful customers, and Gerald's boss had no interest in pursuing the discussion. The farmer, however, became more obnoxious as the conversation went along.

He wasn't the only one getting agitated.

"After they argued for a while, my boss just hauled off and belted him," says Gerald. "That ended the argument."

It was a lesson in public relations that Gerald wasn't sure he'd want to repeat.

Another lesson was in store.

"One of the first and hardest things I learned," he says, "was that you had to be very careful releasing the brakes on boxcars. I went to move a loaded car and I grabbed the wheel the wrong way. When I pressed the brake release, the wheel spun around and threw me off the car. Somehow I managed to shut the engine off, and then I passed out from a concussion. After spending a short time in the hospital, I went back to work and I was always careful after that."

Gerald obviously made a good impression on someone.

"I got a phone call from the Ogilvie grain buyer in the same town. He asked me if I was interested in going to Rosser, Manitoba, as a grain buyer for Ogilvies."

Gerald really had to think about this one. Ultimately, however, he couldn't come up with any good reasons not to go. He remembers exactly the day his career as a grain buyer was carved in stone.

"It was Thursday, May 13, 1948. At the age of twenty-one I was off to my first big job."

Murphy's Law is particularly unforgiving on the first days of a new job. Something could go wrong, and it did. Today rural communities have separate seed cleaning plants, but in those years standard elevators had their own cleaners.

"That first afternoon, a farmer brought a load of wheat in to be cleaned."

Gerald dumped the load into the pit, then ran it up through the cleaner several times, just as he was supposed to. Very quickly, however, he noticed something strange.

"Each time, there seemed to be more barley getting into the wheat."

Gerald was mystified. Where was the barley coming from? He was also concerned. There was no doubt that he was cleaning the grain, but he was also contaminating it by mixing barley with wheat. This could have serious consequences for the customer. Gerald's was a one-man operation and there was no one he could turn to for help. He was entirely responsible. Reluctantly, he confessed to the farmer.

To Gerald's surprise, the man shrugged it off.

"It didn't matter. He was just going to be using it to feed his cattle. Boy, was I relieved! After he left, I crawled down into the cleaner bin and found a bunch of barley stuck in one corner."

That was one episode he wouldn't allow to happen again. Or would he? Once a year, in late winter and early spring, seed cleaning season reared its ugly head. Gerald learned to dread it.

"Someone always wanted barley or oats cleaned during the time you were cleaning wheat," he says. "Someone always seemed to blame you for mixing their seed grain with other grains. Then there was the screening problem."

When seed was cleaned, the leftover screenings were given back to the farmer for use as feed.

"When you only have three screening bins and you clean grain for seven or eight guys in a row, the screenings have to get mixed," says Gerald. "Someone always complained about not getting their own screenings or not getting enough back. I was always glad when seed cleaning season ended."

Next came threshing season.

"In those years there were no big trucks, so many hauled seventy- or eighty-bushel loads. I usually stayed open until the farmers stopped threshing. As I was alone, it meant I had to stay until midnight or one o'clock in the morning entering all those tickets. Then when the farmer came in for a cash ticket, you had to write them all down on another form."

Paperwork wasn't Gerald's forte.

"I was always getting letters bawling me out for not getting my weekend statements in on time. There were many times I sent in two at a time."

He didn't enamour himself to the fire inspector either.

"He used to come around three or four times a year," says Gerald, "and I must admit that I wasn't the neatest person in the world. He was always giving me a blast for having a dirty elevator, or else the fireguard wasn't good enough."

Fortunately, Gerald never had a fire. His troubles lay elsewhere.

"One thing that really annoyed me was trying to open or close some of those doors on those old cars," he says. They were bashed, dented, and warped. Many an agent resorted to a sledgehammer or crowbar to jerk loose the wheels that ran along the doorway track.

Gerald recruited the use of a friend's tractor "just for the purpose of opening and closing the doors." Even then it wasn't easy. One day he simply couldn't get the door closed. A customer offered to help by driving the tractor while Gerald worked on dislodging the door. The man revved the motor, reefing on the chain that ran between the tractor and the boxcar door. Working together, they finally got it closed. Then disaster struck.

"As I went to take the chain off, instead of the farmer backing up to slacken it, he accidentally went ahead. The chain snapped up and hit me in the face. Again I was knocked out."

Hospital visit number two. This time for a whole week.

Gerald wasn't the type to hold grudges. Even when a friend tried to put one over on him, he chose to view the incident with humour. "The guy mixed some treated grain with his good grain, trying to get rid of it," says Gerald. "I didn't notice it in his first load but caught it in his second. He denied it, so I took a sample to my head office in Winnipeg and, sure enough, it was treated. I made him take it back home, and when he came back for his second load he was so mad that he drove about five miles with his box in the air because he forgot to put it down. The funny part about all this is that we were good friends and had spent a lot of time at square-dance parties. To top it all off, he asked me to be an usher at his wedding."

A farmer didn't have to be mad to accidentally leave his truck box up. Bill Jepps of Didsbury, Alberta, recalls the mishap occurring twice in his career as a grain buyer.

"The last load came in for the day and we stopped for a visit, the box of the truck still hoisted," he says. "The farmer jumped in and took off, forgetting his box was up. He took out the end of the driveway, taking the door openers along."

Visiting was common. Elevators in prairie towns often served as a local gathering spot for coffee, gossip, and ongoing card games. Doing business over a bottle was accepted practice for many years. The farmers who hauled to Gerald Hallick's elevator knew how to put a little fun into their work.

"In 1968 we had a really wet year and had to dry a lot of the grain. The Pool brought in a grain dryer, and we dried grain day and night. All the farmers joined together and hauled each other's. One stipulation we made was that each farmer who had [his] grain dried had to bring in one or two bottles of liquor. By the time we finished, we had a great collection of empty bottles. It was fun at the time, but not something I'd want to brag about now. I don't know how many times we plugged the leg and I had to go down into the boot and dig it out. After crawling down those dusty steps and getting down on my hands and knees, I always had visions of a rat crawling up my pant leg."

Surprisingly, for someone who never intended to stick with it, Gerald spent twenty years as a grain buyer.

"There were many times I wished I'd done something else with

74

Note the grain dryer attached to the annex; it is similar to the one the
Pool brought in for Gerald Hallick on a wet year. Forrest, MB
PHOTOGRAPHER/CONTRIBUTOR: DAVE OSTRYZNIUK

my life," he admits, "but I did make many good friends among the
farmers and other people in the grain trade."

Gerald met his wife in Rosser and raised his family there.
Together they became leading members of the community, operat-
ing the general store, driving the school bus, and running the local
post office.

"When I went to Rosser in 1948, I thought I'd only be there a
month, but I'm still here in 2002.

# CHAPTER FOURTEEN

# Snipe Lake

"Our dad never spanked us," says Arlene McDonald, who tells this story in tribute to him. "The boot cleaning was enough."

Wilbur Jose, a man with three lively daughters, had hit upon a good strategy. And it proved increasingly effective as the girls grew. In the mid-1950s teens were just as good at creating their own excitement as they are today, and when things got interesting on a Friday night in the small town of Snipe Lake, Saskatchewan, it often resulted in a boot assignment the following morning.

Arlene laughs heartily in retrospect. "Once it was my oldest sister's turn. She never really did a lot of the things my other sister and I did, but we'd been out late the night before and she happened to be the one to have to clean the boot that Saturday. It's bad enough if you're feeling good, but . . ."

Arlene's father no doubt had a tough time hiding the smile that played about his mouth, but he'd made his point.

Wilbur was one of four elevator agents in Snipe Lake. He ran the Reliance, which was later purchased by Federal Grain. His hands were full with the girls, but he knew how to keep them busy.

"We were more or less raised at the elevator," comments Arlene. "We helped Dad empty the annexes of grain with our clothing taped to our arms and legs to keep it from catching in the augers. Then we'd get in there and clean out the bins."

The girls also coopered cars, but before they could be loaded they had to be moved into place beside the elevator.

"One day Dad had to go to a funeral, so he left my sister and me to cooper a boxcar, and we were to place it at the loading chute. I told her to get on top where you turn the brake, and while I used the car jack, she was to control the brakes."

Everything went smoothly until Arlene's sister realized, too late, that she didn't have the strength to shut off the brake. As she frantically tried to crank the stubborn wheel, the car picked up speed and rolled merrily down the track.

"We overshot the elevator," says Arlene. "Dad had to pull the boxcar back to the right place with a tractor."

It didn't take the girls long to figure out that the grain doors used for coopering made terrific props for their various amusements.

"We took them in the winter time and used them for ski and toboggan jumps," Arlene recalls. "And they were great in the summer. We had a big dugout and used them as rafts. Dad was always looking for his grain doors. He said he had to come and collect them so he could finish his cars."

"It was a great treat to go up the manlift as well," she adds. "We often took other friends to the top of the elevator. We were like monkeys. It was probably much more dangerous than we ever realized, but at the time it didn't seem like it."

The family owned a little fox terrier of whom they were especially fond. Smoky was "black like the night," and a fantastic ratter. He was much in demand by all the elevator agents in town, as he would keep the rodent population in check. One agent in particular relied heavily on the little dog.

"He wouldn't even take a mouse out of a trap," says Arlene. "Many times he came running over for Smoky to help with mice in his office."

Then one day Smoky went missing. They searched and called, but the small terrier was nowhere to be found. After three days, hope wore thin. Devastated, family members went about their daily business with heavy hearts.

The man who was afraid of mice and rats certainly missed the little fellow. Now he was left to face the hideous creatures on his own. Fighting his fear, he descended into the bowels of his elevator to clean the boot.

He'd only been at it a few seconds when every hair on his body prickled. From the eerie depths, a low growl reached his ears. He strained his eyes, and out of the shadows a menacing black form materialized. Before his brain could register that it was Smoky, the agent was up the ladder and out of there. Fortunately his wits soon caught up to him and he realized he'd found the missing dog. He raced over to the Federal to give Wilbur the news.

"Smoky had fallen down the stink hole," states Arlene.

In spite of a joyful reunion, the agent who'd found the dog was horribly shaken, and Smoky was dreadfully in need of a bath.

Smoky and the girls didn't just help at the elevator.

"Dad also had land and farmed from town," says Arlene. "As we got older, I ran the combine, and Mom and my one sister hauled the wheat. The Snipe Lake point was very large, as we were in the heart of the wheat belt.

"It's funny how we take things for granted," she muses. "It seems like at one time we had it all, but it was not to be. As the prairie elevators stood all across the Canadian landscape, we never

Grand opening of the new Saskatchewan
Wheat Pool elevator. Snipe Lake, SK, 1954
PHOTOGRAPHER: UNKNOWN / CONTRIBUTOR: ARLENE MCDONALD

realized that they might one day be knocked down."

As children do, Arlene and her sisters grew up and moved away. She was living in Medicine Hat, Alberta, when Snipe Lake lost its last elevator. On her next trip home, Arlene drove right past the community, missing it completely because the familiar landmarks no longer guided her. Gradually the town deteriorated.

"Like all little towns," she laments, "once the school and the elevators go, there isn't much left. The hall still stands but it's pretty dilapidated. They've built a new part on the end for the Fish and Game, but nobody seems to bother with it. We used to get trains once or twice a day. Now there is only a train every six months or so."

As the businesses of Snipe Lake melted away, so did its people. Today only one person remains, a former resident whose memories of rafts on the dugout and rides on the manlift remain fresh and strong.

"I always tell everybody I'm the mayor," says Arlene, who lives in her parents' original home. "I have an old fire truck that I use to water with, so I tell everyone I'm the fire chief too."

She is quite happy in her tiny domain.

"I have a business for which I travel to different craft shows and trade shows. My business sign proudly displays Snipe Lake as my address. Many people have asked, 'Where is Snipe Lake?' But one day in Regina a lady said, 'I know where Snipe Lake is! My father built the Pool elevator there in 1954.'"

It was a fitting comment, for Arlene's wares reflect the elevators of her past.

"I make bird barns," she explains, "and I put little elevators on them. The elevators are very popular."

She also makes plaques depicting grain elevators. Imprinted across them is a timeless sentiment: "Gone But Not Forgotten."

# CHAPTER FIFTEEN

# My Treehouse; My Castle

Even though Lee Eckstrand grew up on a lonely siding with no friends or neighbours for miles around, he was the happiest kid in the world.

In the 1950s and 1960s his father was a Saskatchewan Wheat Pool agent near Lloydminster, a town sitting on the Alberta/Saskatchewan border. The Eckstrands lived, literally, in the elevator.

"Our cottage was built onto the office," Lee says, "and the elevator was my treehouse, castle, hide-and-seek, exploration, and education centre for years."

It was the perfect playground for a boy who was always reaching for the sky. His father's twin elevators provided him with ample opportunity to climb eighty feet closer to it any time he wanted.

"I loved riding up the manlift," says Lee. "Pops had a five-gallon pail of grain set beside it to be my ballast weight, to make it operable for me. Each year as I grew up, he'd take some grain out so that it would operate easily both ways."

When he wanted a bigger challenge, Lee climbed the ladders to the top. In true kid fashion, he quickly figured out how to keep the dust he stirred up from falling into his eyes. He donned a pair of goggles! Who cared about the rest of him getting dirty?

The most thrilling way of climbing the elevator was scaling the outside. With the help of the lightning rod cable, he often climbed all the way to the roof. There he stood like a monarch, surveying his kingdom. He could even skip over to the next elevator via the adjoining walkway.

"I'd study the countryside for a while, then pop the roof hatch and come down the ladder." No need for goggles this time. With his eyes pointed downward, the dust he raised settled everywhere but on his face.

When the weather was bad, or if he simply felt like it, he stayed indoors. There was plenty to keep him busy.

"I liked climbing up the insides of empty bins, using the corner bracing as ladder steps," he says. He had to feel his way in the dark and occasionally he'd "scare up a pigeon."

Finding a boy in a bin no doubt startled the bird as much as finding a bird in a bin startled Lee. But he never lost his footing and fell.

"Pigeons were one of my main interests in climbing to the top," he says. "Each year when the local summer fair was on, I'd catch two elevator pigeons and enter them. I never got higher than second. Someone always had a beautiful pair of fantails."

Lee and his German Shepherd dog also had ground-level interests. They made a great team keeping the rat population in check.

"The dog loved it when I scared rats out for him to grab. He'd give one quick snap, and then go after another one."

Rats are dangerous, disease-carrying rodents, and pigeons too supposedly carry disease.

"But I lived through it all," Lee says cheerfully. "Never fell, never got sick."

Lee's parents gave their son the freedom to explore and experience, but they also kept a watchful eye, aware of the need to be vigilant raising a child in a potentially hazardous environment.

"The grain company had an annual award for site safety and appearance," he says, "and Pops earned a couple of them. They were his pride."

Lee's parents were fun.

"They often played ball with me in the yard. With the elevators, the goosewalk, the office, and a couple of coal-storage sheds, a sheltered courtyard was created. When I switch-hitted and batted left, I always broke the windows in the elevator by the scales and desk area. Pops would just fix them. He knew there was no bad intent."

When Lee practised alone, the elevator made a great backstop.

"I developed quite a throwing arm and batting technique. As I got older, I could throw over one elevator and find the ball between the two. I could bat even better. The balls would clear both elevators."

Despite the good times they had, the Eckstrands's living conditions were practically primitive.

"Pay was very low then," says Lee, "especially for a low-volume site. There was no running water, no central heating, all the lights were Saskatchewan chandeliers (pull-string porcelain fixtures), and there was no indoor toilet—just a pail."

This was all quite impressive to Lee's friends. When they came to visit, they thought it was "neat because it was so different." Lee, on the other hand, "enjoyed all the amenities at school or at friends' houses."

We never needed air conditioning," he says. "From noon until sundown, we were in the shade of the elevators. It was too cold usually."

Nevertheless Lee was content.

Twin elevators, similar to Lee Eckstrand's "treehouse
and castle." Starbuck, MB
PHOTOGRAPHER: UNKNOWN / CONTRIBUTOR: KATHRYN BARTMANOVICH

"My needs seemed to be slim and my wants even less."

As Lee grew older, he helped in the elevators.

"Dealing with the different farmers was memorable," he comments, "especially at harvest time. They constantly came in to have their grain tested for moisture. Some let me test it. Others demanded that Pops do it—only to get the same reading. Then there were the 6:05 PM farmers who never hauled a bushel until the elevator closed."

Many customers became good friends. Lee remembers with amusement the ones who "stayed and played cards or had lunch, leaving their truck in the elevator while others were waiting behind. But there was no road rage then."

Lee dips into childhood memories with delight and refers to

himself proudly as the "son of an elevator man." Clearly, his parents felt the same connection to this unique way of life.

"When they died," says Lee, "there was a request in both their wills to have the ashes spread where all their pets had been buried: in the fireguard around the elevators."

Lee was determined to honour this wish, but when the time came, it wasn't easy. By then the elevators and fireguard were long gone.

"So it was an estimate," he says, "but I'm sure they're resting in peace."

# CHAPTER SIXTEEN

# Searching

In 1950 Mike Raymer was a small boy who had a big fascination with elevators. He loved it when his stepfather occasionally helped out in the National A in Blaine Lake, Saskatchewan.

"I'd sneak over and watch," he says, "but I couldn't see too much because I was too scared to walk in."

Mike had to content himself with glimpses, fuelled by imagination. The elevators were "a mysterious entity, all dark and spooky and noises and smells."

Four times a day he tickled that imagination as he crossed the tracks on his way to and from school.

"Crossing the tracks meant crossing over the driveway of the elevator," he says, "and I would peer into that dark, eerie entrance. Often I'd see a horse-drawn wagon enter the elevator with a load of grain. I'd catch the aroma of the horses struggling up the incline. In winter I wondered how the horses pulled sleighs over the wooden scale deck. But they did."

Mike spent a lot of time with his cousin. It didn't take long for the elevator agent at the National B to notice the boys' interest.

Painting crew on a Saskatchewan Wheat Pool
elevator. Blaine Lake, SK, 1966
PHOTOGRAPHER/CONTRIBUTOR: MIKE RAYMER

"He was a religious man," says Mike. "Never cursed, never
boozed. He allowed us to loiter around the elevator provided we
didn't get in the way and we didn't get hurt."

In exchange for his kindness, the boys were eager to do odd jobs.

"We swept the receiving scale, and swept the boxcars before he loaded them. He treated us to ice cream and soft drinks, and we got paid a little for our work."

Mike never forgot this generous man and the profound influence he had on them.

"He taught us the value of virtue and morals."

As he got older, Mike was more than ready to take on heavier work. His first job was a lesson in itself.

"To trim the fireguard," he says, "a six-foot perimeter. My cousin undertook the project for a whopping five dollars and I could get half if I did half the work."

Mike's eyes lit up at the prospect.

"Two dollars and fifty cents was a lot of money in the early fifties."

Scythes in hand, the boys dove in with gusto. Very soon, however, their spirits wilted.

"Our eagerness turned to despair," he admits, "when the blisters appeared, as well as the ache to untrained muscles. And we were overwhelmed by the tedium of such a dull job."

It took hours, but they persevered.

"By the end of the day, we were finished and went to get our money, only to be told by the grain buyer that we had to rake up the vegetation we had hacked down. This was a good lesson in working for a living."

That same summer a crew of painters arrived at the National A. Mike couldn't tear his eyes away as they strung ropes and scaffolding for the big job.

"Somehow I developed a friendship with the foreman, who was a really nice guy. He asked me to go to town and buy the men some soft drinks, and I could have a bottle of pop for my work.

"'Sure!' I said. He gave me a dollar, which was enough for six drinks and some change."

Mike hurried off on his errand and was back in no time. He remembers that soft drink with special fondness.

"It sure was good!"

Mike and his cousin continued to work together.

"As we got stronger we could tackle bigger jobs, like coopering cars. At first it took two of us to pick up a grain door. Eventually I coopered cars myself, and with five elevators in this town, jobs were available. I cleaned three boots in one day without any concern for the dirt I got into."

Cleaning boots was a horrible, filthy, necessary evil. They quickly filled with dust, which turned to mud when the boots leaked. The dust or mud mixed with loose, rotting grain, and the stench was unbearable. This foul mixture had to be cleared away once a month or so, in the interests of maintenance and fire safety. But nobody ever wanted to enter that damp, creepy hole. Perhaps that's why boys were often hired to do it. The job required two. One stayed up top, manning a five-gallon pail attached to a rope. He lowered the pail hand-over-hand to his not-so-lucky partner below, who scooped it full, then gave the signal for it to be raised out. The person below soon learned to flatten himself against the narrow wall of the shaft as the pail was repeatedly raised and lowered. One slip from his partner and the whole apparatus could come crashing down on his head. As it was, the pail often banged from side to side on its journey, shaking loose great clouds of grain dust, which settled softly on everything below.

Wade Collinge, who grew up in Richlea, Saskatchewan, cleaned boots with his brother for a couple of the local agents.

"I had the great pleasure of being the guy in the boot," he quips. "We would take all day, as we were young. I remember how soft and thick the dust was. You could run your hand through a ten-inch pile and feel no resistance. I also remember the smell when you got to the bottom. My brother was older and stronger, so he ran the pail. Every time it swung and hit the ladder or the wall, I wore the result. There was no mask. I was always on the lookout for snakes and alligators down in that not-so-roomy hole."

Perhaps watching for alligators was stretching it, but the boot certainly harboured snakes, skunks, rats, and all manner of other creatures. Grown men were known to refuse to enter it for that reason alone.

I wasn't aware of any of that the day Dale phoned me in a state of emergency from his one-man operation in Gwynne. The leg had plugged while elevating grain and he needed an extra pair of hands to unplug it.

"Wear the oldest clothes you've got," he said before hanging up.

I bundled up our three-year-old and down we went. Then down I went (being smaller and weaker than Dale). As I descended I felt as though I was being buried alive. The Gwynne elevator was an Alberta Wheat Pool original, built in the 1920s. The grey, old timbers were blanketed with ancient dust. The air was musty with age. When I reached the bottom, the molding, damp grain under freshly spilled wheat was incredibly offensive to my nostrils.

I looked way up the shaft to two small faces peering down at me and thought, "All they have to do is shut the door and walk away. Nobody would ever know I was here."

Then the light above was obliterated as Dale dropped the five-gallon pail into the opening and I prepared to tackle the job of filling it with the noxious contents of that boot. I lost track of time as I repeatedly filled the pail. It seemed I wasn't even making a dent in the grain. Once, the rope slipped while Dale was hauling up a full load and I watched in horror as the deadly missile hurtled towards me. But Dale's sure hands caught it well before it smacked me into oblivion.

Where Wade Collinge cleaned boots with his brother for a couple
of the local agents. Note the balloon annex, and the boxcars
on the right. Richlea, SK
PHOTOGRAPHER/CONTRIBUTOR: CHRIS STACKHOUSE

I had no company, as far as I know. I'm not afraid of rats, snakes,
or skunks, but show me the most delicate spider and I'm paralyzed.
Thick, dusty cobwebs hung everywhere. I kept my eyes on the grain
and on the pail and nothing else. When I finally surfaced, it was like
being reborn. I had never been so dirty and stinky in my life. I threw
away my clothes, and never before nor since have I more enjoyed a
hot, soapy bubble bath. Boys like Mike and Wade who braved the
boots "without any concern for the dirt they got into" have my hearti-
est admiration.

Of course, *they* didn't do it for free.

"We got two dollars each for all day from one agent," says Wade. "When another agent called and we went, he paid us five dollars each. We thought we were on top of the world! Then came the day the two-dollar guy paid us and he had brand new bills. Two bills were stuck together. Well, my brother and I had quite a dilemma as to whether we should return it or keep it. Sorry to say we kept it and split that dollar. I've felt guilty ever since."

Mike was paid too, and because he was so willing, he was frequently recruited to perform the odd task. One task was very odd indeed.

"I was called to the National A," he says. "The grain buyer was in a dither. He'd lost his spectacles, probably in the grain. He had to run the grain through the cleaner to find his glasses. My job was to watch the cleaner and look into the scalper box to see if they were there. We cleaned grain for two days and still no glasses."

At the end of the second day, Mike headed to the "local billiard academy" to unwind over a game of pool. The owner of the pool hall casually asked whether he knew of anybody who'd lost a pair of glasses. A homeless pair had been sitting there for several days.

The next day Mike was back at the elevator, telling the buyer about the unclaimed glasses. A glimmer of hope sprang into his eyes, but he wasn't taking any chances. He charged Mike to resume his post at the cleaner while he went to investigate.

"He returned shortly," says Mike, "and told me to stop the cleaner. He'd found his glasses."

Good thing, for another day of grain-watching may well have driven Mike to needing glasses!

Locating a pair of eyeglasses in thousands of bushels of grain is akin to finding a needle in a haystack. But Oscar Gudlaugson, who ran the Alberta Wheat Pool in Rycroft, Alberta, had an even tougher challenge. One day a wealthy farmer's fat wallet went into the pit with his grain. By the time the man made the shocking discovery, the wallet and the grain had been scooped up the leg and dumped into an overhead bin. Surely it was gone forever.

"After a long day's work and late supper, Oscar returned to the elevator," says his wife, Fern. "He recycled that whole bin of grain and retrieved everything in the wallet. Paper money, change, and any other items."

It seemed remarkable to have found it all, and even though he was exhausted, Oscar glowed with the satisfaction of his success.

Satisfaction would have to do.

"Reward?" asks Fern. "The farmer brought us two pounds of homemade butter!"

Experiences like Oscar's ultimately discouraged Mike from wanting to be an agent. He loved working in elevators and took a good stab at buying grain. He even trained at the Searle A for three summers, but he readily admits, "dealing with the farmers was a job in itself."

"One particular farmer always drove into the scale room with a cigarette in his mouth. Once, the grain buyer was out and I was in charge. I insisted that he extinguish it."

The farmer looked at the young trainee and then, "being a wise guy," coolly flicked his cigarette into the pit, where it vanished into his flowing grain.

Mike sprang into action.

"I immediately shut the slide at the leg and told him about the danger of fire. He laughed and said not to worry. I pointed to all the No Smoking signs in the driveway. As he finished unloading he drove out, leaving me to find that cigarette."

Terrified that the elevator would catch fire, Mike climbed down into the pit and searched frantically for the cigarette. He came up empty-handed.

"I sat and watched the pit for smoke for the rest of the day. Nothing caught fire, but I started to realize the responsibility required, and all of a sudden my enthusiasm waned. I didn't want to be a grain buyer anymore."

Mike made the difficult decision and spoke to his boss. Fortunately, the man understood. Mike didn't even have to quit his job.

"I continued as a helper but not as a trainee," he says. "A couple of summers later, a repair crew arrived at the McCabe elevator in town. They were short a repairman and asked around to see if anyone was interested. I accepted, and worked for two months with the crew. It was a good job. There were a variety of tasks. I learned as we went along and there were no farmers to contend with."

A whole new world opened up for Mike.

"They returned the next year and hired me again and I got another two months' experience. After all that knowledge, I was able to undertake many jobs myself."

Mike had found his niche. Working on a repair crew was the life for him!

# CHAPTER SEVENTEEN

# A Rough and Ready Bunch

Men on elevator repair crews achieved almost legendary status. Not only were they reputed to be able to fix anything, they seemed downright fearless.

"They had nerves of steel," remarks Verla Nevay, "to be able to get around and work on those high roofs. I just felt sick when I looked up to see one of the crew standing at the edge, looking down to see if I was bringing coffee."

Verla boarded the crew at her house when they came to work on her husband's United Grain Growers elevator in Valley Centre, Saskatchewan. Not all repair crews were so lucky. Art Braun recalls that in Altona, Manitoba, in the 1930s, "they would sleep either in the engine room of the elevator or in the empty flour shed. Bath facilities were often non-existent and they would be away from family for long stretches of time." When my husband, Dale, worked for Agricore, the crews set up in ATCO trailers, centrally situated so that they could attend to several elevators without having to break camp during the one or two weeks they were in the area.

Thirteen-year-old Eugene Ukrainice didn't enjoy the luxury of an

Repair crew at the McCabe elevator: Walter Svetlikoff, Mike
Raymer, and Walter McDonald.
Blaine Lake, SK, 1968
PHOTOGRAPHER: L. POSTNIKOFF / CONTRIBUTOR: MIKE RAYMER

ATCO trailer when he travelled with a repair gang in the mid-1950s.

"My father was the foreman for a crew that covered the prairies
doing repairs or building new elevators for Federal Grain," he says.
"We were doing repairs for the Broad Valley terminal when there was
one hell of a storm. I was sleeping in my uncle's wooden construction
trailer. Wind shoved it that night right up against the elevator. We had
to get pried out of it in the morning. Moved it about a hundred yards.
It was the wildest ride I ever had. It also ripped the new shingles off
the elevator, so we had to put a new batch on the next day."

Eugene worked alongside his father and uncle, but most construction crews left family far behind. That, and the enforced close
company with other men of brawn, may well have been what made
many of them reckless.

"In the days of the wooden elevators, safety was not an issue," says Mike Raymer. "Safety equipment was available but not stressed. We would actually take chances to prove our bravery. Safety belts were for sissies."

Art Braun knows what he means.

"When they worked on their scaffolds without any of today's safety measures, they'd walk about as if they were on the ground. They were a rough and ready bunch who worked hard and also played hard. About 1935 we had a crew in to build an annex. It was terribly hot, so they started work at six in the morning, quit from about noon until five (during which time they cooled off in the local pub), then came back to finish the workday. It always amazed me that after an afternoon of drinking they could be up on the cribbed wall, hammering away with their hatchets and four-inch nails."

Repair crews earned a great deal of respect from both elevator agents and the companies they worked for. Dale recalls that it was always the agent's responsibility to accommodate the men in any way necessary to make their job easier. If a repairman was coming to work on the leg at the bottom of the boot, for example, the boot had to be cleaned first.

"I hated it when something needed to be fixed up in the cupola," says Gerald Hallick, "and it needed two men. It was usually me who climbed that dusty old ladder to the top while the head repairman went up the manlift. One good thing about it was it was always easy coming down!

Mike Raymer points out the downside of such preferential treatment: "As we gained experience and confidence, we would tackle any project the company directed. We got a sense of importance by the success we had with things that seemed impossible. But a sobering experience occurred to bring us to reality. At one audit there was a superintendent present when a boot bearing failed. Two outside repairmen were summoned. They were reluctant to enter the boot because the accumulated dust level was above the boot bearing. Well, the superintendent would have none of that B.S. He berated those repairmen, calling them scavengers and said that working in dirt was part of their job. 'Get the hell in there and change that bearing or else!' he hollered. All of us began to realize how little we meant to the company in terms of importance."

But Mike proved his worth many times. Perhaps never more so than the day he saved an elevator from burning down.

"As the diesel engines in elevators were removed," he explains, "various electrical methods were adopted to power the head drives. One plant had a motor, V belts, and a jackshaft with chain and sprocket system. One windy day when I arrived on site, the grain buyer was busy turning bins. As I walked by the manlift, I detected a wisp of smoke."

Quickly surmising that it came from above, Mike jumped on the lift and shot up to investigate.

"I soon found the source of the smoke. The strong wind blowing on the end wall moved the wall slightly against the jackshaft. The V belts rubbing on the two-by-four grooved the board and with the friction caused smoke."

Mike barrelled back down the manlift, yelling at the agent to shut off the leg, as he ran out the door to get a hatchet and saw.

"I quickly removed the offending two-by-four and brought the board to the attention of the buyer," he says. "Eventually that piece of lumber arrived at head office and a bulletin was sent to all the elevators with the heading: Fire Averted by Alert Repairman."

That wasn't the only near-fire in Mike's experience.

"One winter audit caused some excitement," he relates. "The agent would preheat the diesel engine with a blowtorch to facilitate starting. During the process he was called to the phone and rested the blowtorch against the air cleaner of the engine. By the time he returned, the office was filled with smoke. Being the mechanic, I was sent to see what was the matter. The air cleaner contained an oil filter and had caught fire, emitting smoke. Between installing a new filter and all that heat, needless to say, the engine started quite easily."

Rarely was a repair crew member so available when something went wrong. They had large territories to cover and often branched out singly or in pairs to respond to problems as they occurred. Some problems turned out to be nothing.

Ken Thuen, a maintenance foreman looking after a hundred Saskatchewan Wheat Pool elevators in the 1950s and 1960s, remembers one such incident.

"A guy phoned up and said, 'I can't get any grain to come up the leg. The buckets are going. Everything's working, but it's not taking any grain.' Well, I'd never heard of that, so I had to get down there. It was about a hundred miles. Here he'd started the engine and he didn't get it set right and it was running backwards. The cup belt, everything, was going backwards. That was one of those old diesel engines, and if you didn't get it set at the right point on the flywheel when you started it, it would go backwards."

Another time an agent in Humboldt, Saskatchewan, had a plugged spout leading to his annex. In an attempt to clear it, he took a weight from the manlift and ran it down the wooden spout.

"That thing flew right out of the spout where it was plugged," says Ken, "and crashed through the roof of the annex."

The agent phoned Prince Albert to report the accident. When Ken caught wind of it, he thought he'd better phone back to let the agent know he was on his way.

"I got hold of the operator and said I'd like to place a call to the Wheat Pool elevator at Humboldt," he remembers. "She said, 'Okay.' The guy at the other end answered, and I asked, 'What are you trying to do, running a manlift weight through your annex roof?'

'What are you talking about? I didn't put any manlift weight through the roof.'

'You must have. You phoned.'

'I did not!'

'Oh my God,' broke in the operator. 'I think we have the wrong Humboldt. You're speaking to a grain buyer in Humboldt, Kansas.'"

"To this day," says Ken, "I've never figured that one out."

But Ken had other things on his mind. One early spring day he was sent to investigate an agent's complaint that the leg in his elevator wouldn't move at all. It didn't take Ken long to find the problem. Because of their depth, water seepage into the boots was common, especially during spring thaws. The boot in question had filled right up to the top with water and then, in a sudden cold snap, had frozen solid. The boot cavity was one big block of ice. No wonder the leg wouldn't budge!

"There was no way we could chip that out," says Ken. "We just hung a couple of heat lamps in there. It took about three or four days to thaw. The grain buyer looked after keeping the heat lamps on it."

By then Ken was long gone, off to his next job.

Boots were a big part of a repairman's life. He couldn't be claustrophobic, and he couldn't be afraid of discovering creatures like rats, snakes, or skunks.

"And speaking of skunks," says Ken. "At one place we were putting wooden lathes on the legs to tighten them up so they'd be more dust-proof and oh, man, there was a terrible smell of skunk. We knew they were down in the boot, underneath the pit. One day after dinner I took the .22 rifle from the office and said, 'I'm going to have a look.'"

Ken couldn't believe his eyes when he peered through the trap door opening into the boot.

"The skunk was coming up the ladder. It was straight up and

down and he was climbing it just like a man would."

Without thinking, Ken aimed the gun and fired. The skunk died instantly, tumbling in a heap back to the floor of the boot.

" I wish I hadn't shot him," says Ken. "I should have just let him come up and he would've gone. I guess he didn't want to be down there when we were there anymore than we wanted to be down there with him."

But the deed was done.

When Ken returned to the office and told his partner what he'd done, the man wouldn't believe him. He dashed out to see for himself.

"You lying son of a . . . There's nothing there!" he said.

Ken ran back and looked down the shaft. Not a skunk was in sight. He was dumbfounded. He knew he'd killed it, but where had it gone? Try as he might, he couldn't convince his partner he was telling the truth.

Later that afternoon the man was busy in the boot while Ken worked in the driveway. Suddenly he called up the shaft.

"I guess you were right."

He had shone his flashlight under the pit and there, deep within, saw the skunk's tail. Ken was vindicated.

"There had to be another skunk," he reasons. "Possibly a mate that dragged it under the pit."

Close-up of a leg with inspection door open. Skunks got into the pit and occasionally rode up the leg in one of the cups.
PHOTOGRAPHER/CONTRIBUTOR:
CHRIS STACKHOUSE

That theory proved true for Bill Rempel, an Alberta Pacific agent at Makepeace in the 1960s. When he shot a skunk for the neighbouring Alberta Wheat Pool buyer, he knew it was dead. Blood instantly appeared between its eyes and it fell completely limp. Bill fished it out of the boot with a long piece of wire.

"When the skunk cleared the floor it blinked its eyes," he says. "There was no sign of it being shot and we knew we had a live one by the tail."

They bolted out the door and sent the skunk flying. A few minutes later, they pulled up its partner, "deader than a doornail."

The two agents were lucky they didn't get sprayed. Ken Thuen knows. He once saw a skunk in action. Early one morning he was building doors in the back of the Wheat Pool elevator in Prince Albert, Saskatchewan. It was a beautiful morning, with the sun streaming across the driveway. The grain buyer was busy at the leg, tightening loose cups.

"This particular elevator had an electric motor," says Ken. "There was a switch by the leg inspection door in the driveway. He'd run the leg a bit, maybe ten feet or so, really slowly. He could tell which cups were loose, and he'd stop the belt and tighten them as they appeared."

One cup held a big surprise as it came into view: a skunk, riding the leg like a Ferris wheel! Ken was just walking up when the skunk announced its presence by spraying.

"I could see it," says Ken. "It was just like a mist. You know, like you mist your hair."

As it captured the rays of the sun, the skunk's spray was almost celestially beautiful, but the horrid sting came in milliseconds.

"It got the agent right in the eyes," says Ken. "Oh man, did that burn. It was just like he was blinded for a while."

Ken immediately leapt to the agent's aid, ignoring the skunk.

"He got elevated up and we didn't go up there to find him. I guess he ended up going out in a shipment somewhere."

Verla Nevay's husband, Len, made good use of the leg when dealing with skunks.

"The curious things nosed around in the elevator and usually fell into the pit," she explains. "To get a skunk out, Len elevated it up the leg. It would be sitting in one of the belt cups, and get dumped into a bin, where it was captured and taken away. After its flight up and into the bin, it was a bit woozy, not too smelly, and could be handled quite easily. At Rutland, the Pool agent and Len had skunks digging in under the elevator, so they decided to set traps. But a skunk in a trap so close to the buildings is a rather smelly mess if it is killed there. So Leo and Len tied ropes on the traps and in the early morning walked down the railway tracks, each pulling along their catch. They killed them out of town. No smell!"

Cats were more common and less menacing than skunks. They liked to loiter about elevators, availing themselves of the impressive mouse population. Their natural curiosity, however, tended to get them into prickly situations.

Ken rescued one cat from a metal auger running three hundred feet from a flax elevator to a processing mill. He was testing the auger for proper functioning. It was

Terry the cat poses on the toolbox in the back of Ken Thuen's truck. He travelled with the Saskatchewan Wheat Pool repair crew for seven years. Marengo, SK, 1969
PHOTOGRAPHER: UNKNOWN /
CONTRIBUTOR: KEN THUEN

going full tilt with all the inspection doors open. Somehow the cat got drawn in and tangled in the churning mechanism. Ken raced to save it, sure that it would be maimed or killed. But when he caught up to the cat, it was perfectly all right. It even hung around for a while, enjoying all the attention before continuing on in search of its next adventure.

In Leoville, Saskatchewan, Ken staged another, much more dramatic rescue.

"We were up on the top floor, putting belting on the ends of spouts to deflect the grain from always hitting one spot. I was sure I heard a cat meowing."

"No way is there a cat up here," said Ken's partner. But Ken was convinced.

"We put a trouble light down into the bin where I thought I'd heard the meowing and sure enough there was a black cat down there. He must have got elevated up the leg."

"I'm going down there," he announced as he dropped a rope into the bin.

"You'd better be careful," warned his partner. "That cat's liable to be wild. You could get hurt."

Ken wasn't listening. He descended the rope into the bin and approached the animal.

"He was a huge black Tom," says Ken, "and as friendly as could be."

The cat obviously recognized its saviour.

"I lifted him up and he crawled right around my shoulders."

With the cat clinging securely to his neck, Ken climbed the corner braces out of the bin and came down to ground level. He figured that the cat had been trapped in the bin for several days, and he wasted no time in finding it some milk.

"It's amazing that he wasn't injured," says Ken. "He had to have gone right up the leg, into the head, and down and out the distribution spout. He probably would've starved to death in the bin."

For the remaining two weeks that Ken and his colleague were in Leoville, the cat became their third partner.

"When we left," says Ken, "a farmer came and took the cat, so he got himself a good home. I was very happy about that."

One cat that Ken found decided to stay. It happened when the repair crew were working in Marengo, Saskatchewan. They'd gone to Kindersley to do some shopping, and while walking down the street noticed a big grey Tom. The cat was handsome, but it was his behaviour that caught their attention. He was at an intersection, and he seemed unusually streetwise.

"Kindersley had a couple of sets of traffic lights at that time," explains Ken. "The cat stopped at the red light, and then as soon as it turned green, away he went to the other side."

As the men watched the cat, they realized it must be a stray.

"We had some hamburger," says Ken, "so we put it in the truck. The cat jumped in. We came back after a while, and there he was, sleeping in the truck."

For the next seven years the cat never left the repair crew. They named him Terry and he travelled with them to every job.

"He knew the elevators like the back of your hand," says Ken. "It didn't matter to him which elevator it was, he started hunting right away. When we left for the weekends, he'd sleep under the trailer. It's funny we never lost him."

But Terry couldn't be lost. In Leask, Saskatchewan, the agent didn't associate Terry's appearance with that of the repair crew.

"He saw him and thought he was beautiful," says Ken. The next thing Ken knew, the man was "running across the tracks with the cat."

"He was taking him home," he says. "When the agent came back

I asked, 'You didn't happen to see a nice big cat around here, did you? It's mine.'"

Another time, Terry had the ride of his life.

"He always climbed up into the motors of trucks to stay warm," says Ken. One day he failed to climb down before the driver took off.

"That farmer lived about six or seven miles away. Terry rode in his motor the whole way. Then he jumped out and walked back to the trailer."

He was uninjured and unruffled. It was just another interesting experience to add to his growing collection.

When Ken left the Saskatchewan Wheat Pool in 1969, Terry went with him. He spent a cozy retirement as a house cat, his elevator days a thing of the past. He still had a long life ahead of him, though.

"We had him for about eighteen years," says Ken.

Terry was no doubt missed by all who'd come to know him well. Many grain buyers liked a cat or two around the premises. They were good for keeping down the mice. The odd cat was even brave enough to hunt the rats prevalent in Saskatchewan and Manitoba elevators. Alberta was supposedly rat-free, but more about that later.

Agents made no bones about hating rats. Mike Raymer recalls once when he had to think fast just to keep the extra pair of hands he'd recruited.

"I was sent to a diesel drive plant to replace the main drive belt that had derailed off the pulleys and jammed under the receiving scale. I couldn't free it, so I asked the grain buyer to help me. He admitted that he was scared of rats and wasn't about to climb in with them. I assured him that there were no rats. I was under the scale already and didn't see any. Reluctantly he donned his coveralls and joined me, pulling on the belt. In the semi-darkness something ran past his leg."

The agent's reaction was instantaneous.

"A rat!" he yelled. "I'm getting the hell out of here!"

"No! No! It's a cat!" hastened Mike. "Can't you see it? Look, there it is."

Fortunately the agent believed him, even if Mike didn't believe it himself. He stayed to finish the job.

Eugene Ukrainice certainly wasn't afraid of rats when he helped his father and uncle on a repair crew in Manitoba.

"I recall one location where we played slap the rat," he says. "Floors in old elevators were laid out on railroad ties. When repair work consisted of replacing the floor, about a half-dozen boards were taken off the sides from the bottom up. After the floor was removed, there were just ties left. Under the ties were many rats. Everyone tied his pant legs up at the bottom and grabbed a two-by-four. One of the guys had a big pry bar and started lifting ties. Naturally the rats started running for the hills. We had about six guys swinging two-by-fours like Babe Ruth. Rats never got up our pant legs, but the shins took a beating when someone missed the rat, or over-swung, nailing someone else on the follow-through."

Rats notwithstanding, working on elevators had its advantages. From the roof the views were spectacular. While re-shingling an elevator in another location, Eugene's uncle "spotted a big orchard of wild plums just to the northwest of the terminal."

"They sure looked great," remembers Eugene, "and that evening after supper we went to pick some."

They didn't consider that they were on private property.

"We got a bucketful apiece when some guy hollered, 'You're trespassing! Get the hell out of here!' and he fired a shotgun at us. You've never seen an old man and a young kid fly over a barbed wire fence so fast."

The two held their breath, waiting for the inevitable fallout.

Mike Raymer on his last day of work before the high throughout
facility was sold. "My career has ended." Blaine Lake, SK, 2002
PHOTOGRAPHER: STEVE SPEARMAN / CONTRIBUTOR: MIKE RAYMER

When none came, they sighed with relief, and went on their way.
But the incident didn't leave Eugene entirely unscathed. He's never
eaten a wild plum since.

As wooden elevators started coming down in favour of larger

and fewer concrete terminals, repair crews were needed less and less. There were mass layoffs.

"In 1986 a multinational grain company started construction of a concrete and steel elevator virtually across the street from my residence in Blaine Lake, Saskatchewan," says Mike Raymer, who by then was semi-retired. "I was hired as drywaller and painter for the office and exterior painting. During the many visits by company officers, I was told that this elevator would require no maintenance because it was brand new with state-of-the-art electronics."

After all his years of experience, Mike knew better, but there was no sense arguing. Soon enough he was proven correct.

"During the first six months of operation, a strange noise was heard from the head drive. I convinced the operator that something was amiss, and he hired me to go atop and investigate. Taking off the cover of the pillow block, I discovered that the bearings were without lubricant and were wearing from the extreme friction."

This information threw everyone into a flap. Who was responsible for mechanical problems on a maintenance-free elevator?

"The operator notified the company," says Mike. "The company notified the builder, who in turn notified the supplier, who notified the manufacturer. After much discussion, the onus was on the builder to replace the bearings because he was supplying the warranty. The new bearings were installed fully lubricated. After this episode, the grain company hired me as maintenance and repair man to prevent breakdowns."

Mike got the last laugh for repair crews everywhere.

It would've been nice to end on that note, but change is merciless.

"In June of 2002," says Mike, "that elevator is empty and has been sold to a private individual for purposes unknown. All of the surrounding elevators have been destroyed. My career has ended."

# CHAPTER EIGHTEEN

# A Life That Owned You

Joe Storey had a laughing wheelbarrow. Just ask anyone who knew him in Gordon, Manitoba, where he ran the Manitoba Wheat Pool for seventeen years.

"The Storeys often entertained their grandchildren through the summers," says Cathy Picken, who years later married Robert, one of those children. "You often saw Joe pushing his wheelbarrow to the elevator, with the sound of laughter coming from the box."

Grandpa's elevator was a favourite haunt for Robert. Especially after he became the proud owner of a single-shot pellet gun. Pigeons were a menace in prairie elevators, and it was common to keep them at bay by shooting them. Robert got lots of practice on pigeon patrol and his aim got mighty good.

"One day Joe bet him a dollar that he couldn't hit the light bulb some sixty feet above their heads," says Cathy.

What a mistake for both of them. Robert took careful aim and picked the light off with ease. Somewhat chagrined, Joe not only had to part with his dollar, he also had to climb that height to change the bulb. Robert was enormously pleased with himself. He doubled over with laughter at the sight of his grandfather climbing to change the bulb. Soon, however, he was doubling over from something else.

"A dollar bought a lot of candy," says Cathy. "Needless to say, Rob had a tummy ache that night."

When things picked up at the elevator, Joe was all business. Trucks lined up from morning until night.

"They were twelve deep, waiting patiently for their turn at the scales," says Cathy. "On one of his busiest days, they put through over sixty trucks."

For five years Joe had a helper from Gros Isle. Together they could load "more than a dozen rail cars in a busy day." Michael, however, occasionally arrived for work in need of a little jump-start.

"Joe affectionately recalls having to send him up to the house for coffee because he wasn't going to be much help until he did."

Joe's wife Elsie was terrific. Her house, graced by "the most colourful garden in the area" was a stone's throw from the elevator.

"She always had coffee on for anyone who would drop by," says Cathy. "Joe and Elsie loved their neighbours and were often among fourteen other couples who annually brought in the New Year together."

It was a good life.

"A life that owned you," says Cathy. "You could remove yourself from it, but you never really forgot. If you worked the elevators, you knew everyone in the area, and they knew you. Respect was there because you had to be honest and fair in all your dealings. Your integrity was your reputation."

It's been almost twenty-five years since Joe retired. Living in Kelowna, British Columbia, he doesn't see many elevators any-more. It doesn't matter. Their essence lives within him.

"According to him," says Cathy, "there is nothing better than stopping by the side of the road on a sunny fall afternoon and watching the farmers combine. He says if you close your eyes and breathe deeply, you can travel back in time and still recall the taste and smell of the elevators and the grain dust."

# CHAPTER NINETEEN

# Moving On

In the seventeen years that Keith Robins bought grain, he worked for two elevator companies and moved his family seven times in twelve years across three provinces. In this he was not unusual. Elevator agents were continually on the move, whether for reasons of economy, ecology, or simply climbing the ladder. Dale and I, along with our daughter Brandi, moved eight times in twenty years while Dale worked in nine different points. Over time, children of elevator agents dubbed themselves elevator brats, akin to the army-brat label attached to children of military personnel.

"Our daughter started grade four in her fourth new school," says Keith's wife, Marlene.

Many would expound upon the disadvantages of such a lifestyle, but ask former 'brats' what they think and the answer might surprise. Most, like their parents, appreciated the experience of so many places and people. If anything, the damage lay in becoming accustomed to moving regularly, consequently having difficulty settling down as an adult. It is of course sad to leave special people behind, but one quickly learns that friendship transcends time and space.

Millwood, MB, 1959–62
PHOTOGRAPHER: UNKNOWN /
CONTRIBUTOR: MARLENE ROBINS

Glossop Siding, MB, 1962–67
PHOTOGRAPHER: UNKNOWN /
CONTRIBUTOR: MARLENE ROBINS

"We met many wonderful people and still keep in contact with them," says Marlene. "Several of them even got together with a surprise visit and meal for our twenty-fifth wedding anniversary many years after we left the place."

The place she refers to is Millwood, where she and Keith transferred with Manitoba Wheat Pool shortly after their marriage in the late 1950s.

"Our furniture was in the house there before we ever saw where we were going to live. We had no money to go see the place. We just thought it was a good thing to do: move to Millwood in the Russell, Manitoba, area."

After three wonderful years they transferred again, this time with two preschoolers in tow. Glossop Siding was situated "right in the middle between Strathclair and Newdale, Manitoba." It turned out to be their most long-standing location. After five years there, Keith left the Manitoba Wheat Pool for Pioneer Grain. They were on the move again. A whirlwind two years in Alberta saw them first at Elk Island Siding

Moving a Saskatchewan Wheat Pool elevator through the tiny village of Herschel. Note the shovels still hanging on the wall, and the office and drive shed coming along behind. Herschel, SK, early 1960s
PHOTOGRAPHER: KEN MCLACHLAN / CONTRIBUTOR: ELIZABETH MCLACHLAN

near Edmonton and then in Thorsby. They were just getting used to life in Alberta when tragedy struck, putting an abrupt halt to both their grain buying days and their wandering lifestyle.

"In 1969 my mother passed away at the early age of fifty-eighty," says Marlene. "We came back to Manitoba to sit on the other side of the elevator agent's desk—as farmers, back on the farm where I was born and raised.

Marlene will never forget her years as a grain buyer's wife.

"In the grain cleaning season I took meals to the elevator for my husband and also any farmer who was there. I know about flax liners, coopering cars, and manlifts."

But her fondest memories are of people.

"Being wakened at three in the morning by our new-found friends and neighbours, just for coffee and a visit," she reminisces. "Making lunch in the early hours of the morning for curlers returning from the Farmers Bonspiel, in which the elevator agents were always included.

"There are none of Keith's elevators left now. They have all met with some bad thing. Closures and fires mostly."

# CHAPTER TWENTY

# Snookie

My husband, Dale, along with his seven brothers and several cousins, had a lot of fun growing up in Herschel, Saskatchewan, in the 1960s. A big part of village life centered on the many elevators that made Herschel a thriving agricultural point. Pete Cormack of the UGG and Snookie Sanderson of the Federal were friendly rivals. Their elevators stood back to back across a double row of tracks, literally a stone's throw from one another. Large trackside doors, each with a window of four small panes, slid open onto a platform for loading cars. It's unclear who started it, but a slow spell in any given day was sure to find either Pete or Snookie standing on the platform, a pile of good-sized stones at hand, taking careful aim and trying his level best to hit the windows of his competition.

Snookie was a man of character. Anyone who'd pulled off what he did during the Second World War had to be. From start to finish, he'd been on the front lines, working alone as a dispatch rider carrying messages through enemy territory. More times than he cared to remember, he'd ridden a thin and terrifying line between life and death.

Years later he stepped out of his elevator office one day to see Dale with his cousin Raymond practising maneuvers on their new motorbikes.

"Used to do a bit of riding myself," he called out. "Mind if I have a go?"

The boys were taught to respect their elders. Raymond dutifully handed over his bike. The teens couldn't believe it when a man they considered old mounted that little Suzuki. They stood bug-eyed and drop-jawed as he tore off down the road, popping a wheelie for good measure. Then he suddenly stopped, spun the bike around on the gravel, and headed full tilt for the ditch. The boys were just about to cover their eyes when Snookie launched the bike like a rocket and sailed over the gully, clearing it with inches to spare. Then he opened throttle and raced full bore across the pasture, bouncing over rocks, ruts, and gopher holes with ease. Later he laid planks from the ground up to the wooden handrail on both sides of the bridge spanning the creek. He showed the boys how he could drive with precision up one plank, across the narrow rail and down the other side. Dale and Raymond were awestruck.

"All things I did in Italy during the war," Snookie casually announced. "I had to learn some pretty fancy moves to make it through enemy lines and still keep my hide. If you ever want to *really* learn to ride those bikes, I'll teach you."

The boys fairly burst with admiration and pride.

When he wasn't playing with rocks and motorbikes, Snookie was an excellent grain buyer. He had many friends and was well respected in the community. By the time he came to Herschel he didn't indulge, but like Gerald Hallick and so many others of the day, he wasn't opposed to doing business the old way. Whether farmers brought their own or the agent reached into the bottom drawer of his desk, there was usually a bottle on hand to ensure a congenial relationship between agent and farmer. In elevators everywhere, Friday afternoons seemed specially suited for this brand of public relations.

Beneath the office floor of Snookie's elevator was a steam engine, beside which was the water tank that fed it. Snookie checked the water level in the tank by peering into it through a hole in the office floor. It was a small hole. About whisky bottle size.

How convenient!

As soon as a whisky bottle was emptied, it was slipped through the hole to sink into the water. No mess. No bother.

Eventually Snookie's elevator was upgraded from steam to electricity. Imagine the maintenance crew's surprise when they emptied the water tank to discover it was three-quarters full of whisky bottles!

As prevalent as imbibing at the elevator was, it goes without saying that some wives didn't care for it. This posed a knotty problem. Where there's a will there's a way, however. In another small town the grain buyer had an arrangement with the railway agent down the track. When the coast was clear, he'd dial over and repeat a certain message about boxcars. To the untrained ear it sounded like business, but it was really a code for the station agent to bring over a case of beer.

Today such antics are strictly prohibited. Snookie has passed on, and so have the railroad and most of the elevators that once dominated the tiny village of Herschel. Only one remains, reduced to being a gathering station for a larger terminal in Kindersley.

Like too many other Saskatchewan towns, Herschel has lost its grocery stores, farm implement dealership, service station, lumberyard, bank, and school. But the people of the area refuse to let their village die with the elevators. Their soil is rich in more than just wheat. The discovery of three dinosaur skeletons, tipi rings, petroglyphs, and a buffalo jump have spurred the development of the

The Valley Centre UGG on its way to a new location near Herschel,
Saskatchewan. Herschel, SK, early 1960s
PHOTOGRAPHER: KEN MCLACHLAN / CONTRIBUTOR: ELIZABETH MCLACHLAN

Ancient Echoes Interpretive Centre, a popular tourist draw. Among
its many attractions is a gallery housing seventeen original paint-
ings, which depict the relationship of the buffalo to the Plains
Indians who once inhabited the area. It's the largest collection
under one roof of Manitoba-born artist Jo Cooper, an internation-
ally recognized Metis painter and photographer. In true prairie
fashion, the Centre is run exclusively by volunteers. So how could
they afford the 20 thousand dollar collection? One of Herschel's
doomed elevators was the key. It was hauled out to a pasture and
dismantled, board by board, nail by nail. The sale of the wood
raised almost 15 thousand dollars for the Interpretive Centre: a
unique and gratifying swan song for Herschel's prairie giant.

# CHAPTER TWENTY-ONE

# Dam(n) Progress

Like Herschel, Dunblane is no longer the thriving community it once was. Its population and businesses have dried up and blown off across the Saskatchewan prairie. But Len Bartzen, born there in the early 1930s, remembers its heyday. He took over the Pioneer elevators in 1961, in the midst of Dunblane's biggest and final boom, when the town swelled to six hundred people during the nine-year construction of nearby Gardiner Dam.

Len didn't have a clue about buying grain when he stepped into the job. His only experience was helping the outgoing agent with his cut-off, a complete inventory of all the grain and other products on site. Cut-offs took place every time a new agent came in, and/or every three years.

"I worked with the past buyer for four days while we made the cut-off," says Len. "It ended on a Friday night. Monday morning I was the one and only man there to buy that load of wheat that came in about 9 AM. A feeling of 'What will I do first?' hit like a ton of bricks."

The two Pioneers were old buildings.

"There were no dust collectors," says Len. "And only a low roof on one driveway, so we had to shovel about half the grain off, as we couldn't lift trucks high enough."

In addition to that, the air hoists in the elevators leaked. By morning, they wouldn't work, so Len had to run "the old Fairbanks-Morse Z-type engine" to pump air into them before he could take any grain.

"I suppose God came to give me a little hand that first morning," he says, "as the farmer who appeared was one of the few with a hoist on his truck."

Len soon caught on to the job.

"I usually did all my own coopering and loading of boxcars. We had no car mover, so it meant jacking up the cars by hand and jacking them out of the way so we could get more cars in. Later I bought an old tractor and that was quite a help. In the winter a lot of times the track drifted in with snow and had to be shovelled out by hand. Nothing was simple or easy in those days."

Rats lived in the boot, which was trouble enough, but in the spring it got worse.

"The boot leaked, drowning the rats that had buried themselves in the dust. One of the worst plug-ups I ever had was on such an occasion."

Len couldn't unplug the boot alone so he called on his Saskatchewan Wheat Pool neighbour for assistance. The man stayed up at the opening to the boot, manning the rope with the pail.

"I was down in the dust, mud, and rotten dead rats," says Len, "filling the pails. Working with my bare hands, and the stench of those dead rats, all pulling apart, was very hard for my stomach to take. Every time I bent over to fill the pails, I gagged."

Viewing the scene from above, the Saskatchewan Wheat Pool agent couldn't help but laugh. That made Len laugh too, albeit not so heartily as his colleague.

Even if the boot wasn't plugged, elevator agents had no choice but to descend it. No one ever knew when the fire inspector would show up, and the boot had to be kept clean to maintain safety standards. It was a losing battle.

"Those old plants leaked so much dust out of the elevating leg that it was impossible to keep the dust down," says Len. "But you always had to make sure that your plants were ready for inspection because you really didn't want to get demerit marks. It meant your job if you got a certain number. The visit from the fire inspector was a nightmare, but I never did get any demerits. Just a couple of warnings. I guess we were always lucky and had the plants cleaned up just in time."

Len had only a grade school education but his creativity, willingness to work, and acute business sense ensured his success.

"I learned a lot about grain buying, business, and accumulating customers with extra services. When I started with Pioneer, the Pool had about 85 per cent of the business. By the third year I had about 65 per cent. After about four years, the Sask Wheat Pool in Dunblane decided to sell its elevator to Pioneer. So then I had three elevators."

Len welcomed the challenge. It simply meant the opportunity to expand the service he offered. He ran the elevators efficiently.

"I tried to fill up two out of three plants and only keep one for use throughout the main winter months," he relates. "If I needed to load cars from the other two, I knew in advance and started the fire in the office the night before."

Len's philosophy was simple: what was good for the farmer was ultimately good for him.

"We stayed home on long weekends in harvest so that people could get moisture tests," he says. "I felt it was better that they didn't harvest tough grain if I could help it. I had to buy it later."

Len also opened an hour earlier during grain cleaning season.

"That really was a chance to build business. I changed my cleaner many times a day just to keep people in the fields. I was cleaning grain for people forty miles or more away. As a result, they started hauling their grain past their own elevator to me. I once commented to one or two who hauled an extra seventeen miles to me that it was nice, but maybe uneconomical."

"'Len,' they said, 'You kept us in our fields by working so much time that you didn't get paid for. The least we can do is give you our grain business.'"

Len went beyond the call of duty in other ways, as well. It was part of his job to sell wood and coal, along with farm chemicals and fertilizer and the equipment to spread them with. But he also sold motorcycles, snowmobiles, farm workshop equipment, and insurance. Eventually, Pioneer updated his facilities, giving him more grain handling capacity. Len Bartzen was going strong even as Dunblane's boom rounded the corner and began its fatal decline. As construction of the Gardiner Dam ended, workers who'd lived with their families in Dunblane for nearly a decade began to leave. Worse, the economic benefits of the dam itself proved useless, and even detrimental, to the community. Dunblane's hilly terrain was not conducive to the irrigation the dam offered. What's more, its location severed several farmers from the trading area. Len's biggest venture took place as he attempted to staunch the bleeding of business from the community.

"Our local Chinese restaurant/grocery store closed up, and I decided that our little town needed a store," he says. "So I bought a pop dispenser for out in the elevator porch (twenty-four-hour service), and filled the basement of my remodeled office with storage shelves. I built some shelves in the office itself so people could shop. We sold milk, cream, and butter, so I had to have a fridge. And deep-freeze too. We had meat cut at the packers and we wrapped it and

Grain buyers were generally quite involved in their communities.
Thomas Small, UGG agent in Lamont, Alberta,
gets ready for a parade. Lamont, AB, 1965
PHOTOGRAPHER: UNKNOWN / CONTRIBUTOR: ADDIE LEVY

sold it frozen. We had a good variety of canned goods and got in fresh greens once a week as well as milk and bread. People really supported that little store. They knew what it was like when the restaurant closed and they couldn't get soft drinks, chocolate bars, milk, and bread. I did have to put an air conditioner in the office as it was too hot in the summer to keep chocolate bars."

If it weren't for his wife, Grace, and their family, Len never could have accomplished what he did. He spent long hours away from home, but they stood behind him all the way.

"They went to the store sometimes at night, if a customer called and asked if they could get something after hours. Grace was the village secretary, so we moved the village books into the office. That was a big help to me, as she could sell groceries if I was at one of the other elevators or loading cars. We were also the CNR freight drop off and CNR money order office. That worked well, since we were the only business in town anyway."

But by 1969 the town was officially dead. The village council, of which Len was a member, wrapped up its affairs. Along with a few others, Len stayed on, but it would have been irresponsible to ignore the call of opportunity elsewhere. The dam would not defeat him as it had the town. In 1975 he left for Outlook to set up a company for Oliver Irrigation of Lethbridge, Alberta. It was the first of several successful business ventures.

"I've never regretted my days of hard work and learning buying grain," he says. "Those fourteen years were the best business schooling I ever had."

Today, at seventy, Len still runs his own business and has many other irons in the fire. It can't be said, however, that he never looked back. In 1993 he and Grace organized a heartwarmingly successful Dunblane reunion. Of the seven hundred former residents they tracked down all over North America, half made the trip to attend. The reunion was held in a hotel in Saskatoon, because little more than ruins remained of the town. Even as Len and Grace excitedly made plans, the Dunblane elevators closed for the last time. The derelict buildings stood abandoned for four years before they were finally demolished.

# CHAPTER TWENTY-TWO

# Two Bills

Bill Rempel stared hard at the checkerboard. He didn't expect to win. He never did. Why should today be any different? Frankly he'd rather be out loading cars—if there had been any to load.

When Bill began with Alberta Pacific Grain in 1960, he had no idea that playing checkers was a requirement of the job. His boss, however, was "nuts about the game, and the board was always ready and waiting."

"Many of his customers came in just to play a game or two," says Bill. "To keep himself sharp and in practice, I had to play him whenever time permitted. He beat me soundly every game."

But this time, as Bill stared at the board, a pattern began to emerge. He looked harder, and with a sudden flash of insight caught on to his boss's strategy.

"It was like a revelation to me," he says. "I blocked his move, and he never beat me again."

Strange to say, "he very seldom wanted to play after that."

In elevator offices throughout the prairies, many other games were played. It is joked that the game of cribbage was invented in the

Jefferson, Alberta, elevator. Rip-roaring ball games took place in Falher, Alberta, with a metre-stick bat and a Buctril M air freshener ball. In Richlea, Saskatchewan, Wade Collinge's father, Roy, remembers the agent who ran a barbering service out of his elevator on Sundays.

"He cut hair by the old clipper way. Lots of business at thirty-five cents a head. Afterwards a poker game started. That's where I lost my shirt one Sunday. But I got lucky the next Sunday and won it all back. That's when I decided to quit."

Bill Jepps was another agent who enjoyed cards.

"Almost every morning there was a smear game going on in the Alberta Wheat Pool office," he says. "When there were cars to load,

Acadia Valley Alberta Wheat Pool—now a teahouse and working grain museum. Acadia Valley, AB
PHOTOGRAPHER/CONTRIBUTOR: CHRIS STACKHOUSE

my second man and I quit to go to work, but the game kept on with the other players. A lot of the regulars soon got stuck with nicknames. Being the agent, I was The Proprietor. Ben was Cousin Weak Eyes (poor sight). Then there was Mac the Knife, Old Yuk-Yuk (the way he chuckled), These-cards-hate-me Jack, and Hickey-time Bill. Loads of fun, and some bad language too."

Bill Jepps readily admits to the rough language that often flourished in elevator environments. The women who worked as secretaries in those offices deserve huge credit. They were alone in a world dominated by men, many of whom shed their gentlemanly ways to get on with the business of feeding the world. Every one of those women held their own and some even dished back what they got.

Bill Rempel's checker-playing days were numbered. In time he was transferred to Makepeace, halfway between Bassano and Hussar. The Alberta Pacific elevator there was old, with a rope-driven leg, run by a Case engine, in the basement of its office.

"Every time the leg had to be used," says Bill, "I had to run downstairs and engage the clutch. The company refused to install a clutch in the driveway."

There were other vexations too.

"I had quite a few plug-ups as a result of my impatience when I tried to feed the elevator too fast."

Bill's greatest distress, however, came as a complete surprise under the elevator one day. He had wedged himself into the narrow space beneath the back pit to grease the auger that ran the length of the elevator and its two annexes. It was a routine task.

"I had to slide on my belly along the auger box," he says, "pushing the grease gun and flashlight ahead of me."

While in this vulnerable position, he suddenly heard a screech right beside his head. He lifted his face, peering around him in the flashlight's halo. He heard it. Then he saw it. Then he felt it.

The weasel launched itself straight at Bill.

"He jumped over my head and face, and landed in the middle of my back."

Bill was riveted. It was well-known that weasels could be vicious. In his position, Bill could neither retreat quickly nor defend himself. Fortunately, the weasel had no intentions of sticking around.

"I couldn't see him leave," says Bill. "But I was sure relieved he hadn't attacked my face. There would've been nothing I could do."

Alberta Pacific eventually became Federal Grain and after five years in Makepeace, Bill Rempel was transferred to Didsbury, Alberta. In 1972 the Alberta Wheat Pool purchased Federal, and both Bill Rempel and Bill Jepps were given added responsibilities in their respective communities. Bill Jepps inherited an old Federal elevator.

"I got an extra fifteen dollars to run that old pig," he says.

In Didsbury, Bill Rempel was put in charge of "all the Pool facilities in town."

"This was three elevators and it became too much for me," he admits. "I caved in and was hospitalized for a time."

Medical tests revealed that he was allergic to grain dust! In 1975 he wisely left the business for good.

Bill Jepps remained with the Alberta Wheat Pool a total of thirty-five years, spending the last twenty-one in the same station.

"As in most small villages," he relates, "the grain buyer, once he is known and assessed by the community, ends up being part of the village council, fire brigade, Legion, and church council. As a village councilman, I was automatically on the hall board, was deputy mayor, and for the last months I was there, mayor. I was our Legion secretary and treasurer for seventeen years, and on and on."

Almost in the same breath, the two Bills say, "All the elevators I operated are now gone."

"In my case," says Bill Jepps, "that was five or six. One elevator burned down a couple of years after I left. It was rebuilt in 1968, and when the rail line was abandoned, this elevator was sold to the village for one dollar. It is still standing in Acadia Valley, Alberta."

The office has been converted to a tea room/gift shop, and the main plant to a working grain museum. Each year hundreds of visitors stop by to tour the facility and learn how a prairie elevator really operated.

Perhaps more significant is the elevator's continuing presence as a social gathering spot for the community. As volunteer Pat Didychuk states in the Spring 2001 issue of *Legacy* magazine, "Anything that closes in a small town takes away people and the spirit of the place. [Saving the elevator] is so worthwhile for the community . . . and it's a centre, particularly for the seniors. They love to come up to the museum and have a cup of tea and a piece of pie."

# Trouble of All Kinds

In the late 1940s a frightened but determined farmer placed his shotgun on the seat beside him and drove a wagonload of wheat towards Alliance, Alberta. He never made it. An angry knot of men patrolling the road accosted him, unhitched his horses, and dumped his wheat on the ground. Nearby, a group of teenage boys heard the commotion and scurried over to investigate.

"Get lost!" shouted the men. "Someone could get hurt."

The boys immediately backed off.

The impoverished farmer felt shamed and crestfallen. He'd only wanted, needed, to feed his family. Thirty bushels of wheat hauled to the local elevator seemed harmless. But it wasn't. In the midst of a farmers' strike, the delivery of any kind of produce, whether eggs, milk, cream, or grain, was strictly prohibited.

Leonard Wold was one of the boys who witnessed the incident.

"No one was hurt," he says, "but there were some very tense moments."

The event stayed with him because Leonard himself was from the farm. A few short years later he was farming full-time alongside his father, on the land his grandfather had broken and homesteaded.

By 1969 grain was hauled exclusively by truck. When space was limited, there was always a mad scramble to haul as much as possible before the elevator became full, and plugged. At such times Leonard and his neighbour, Norman Brynland, worked together.

"He hauled to United Grain Growers," says Leonard, "so when the UGG had room I helped him haul and when the Alberta Wheat Pool had room he helped me haul. One Friday our Pool agent, Jim Furnald, announced that he would have room on Monday morning and would open the doors at 8 AM."

Both Leonard and Norman knew that would mean a long lineup waiting at the elevator doors well before eight o'clock on Monday. They also knew that if they wanted a chance at that space, they'd better be in the lineup, preferably as close to the front as possible.

"We agreed that some farmers were earlier risers than we were," says Leonard. "So we decided to load our two trucks Sunday afternoon and park them in front of the elevator doors so we'd be first in line Monday morning."

Leonard's fiancée, Mary, a teacher in Calgary, was visiting for the weekend. Leonard had her follow him to the elevator in her car. After he parked his truck in front of the doors, he hopped in with Mary and they drove back to the farm. Monday morning Mary returned him to the elevator on her way through Alliance back to Calgary. The plan was beautiful, for by the time Mary dropped him off Monday morning, there were twenty-four grain trucks lined up behind the two that Leonard and Norman had planted there the day before. Leonard said goodbye to Mary and strolled towards his truck, no doubt congratulating himself on his fine thinking.

"At that time we weren't in the habit of taking keys out of vehicles," he says.

It would be a simple matter of jumping into the truck, firing it

Elevator row. Alliance, AB
PHOTOGRAPHER/CONTRIBUTOR: CHRIS STACKHOUSE

up, and being the first to drive into the elevator as soon as Jim opened the doors.

"Suddenly it struck me," says Leonard, "that at the last minute on Sunday I'd decided to take the keys out and throw them on the dash of Mary's car. Oh boy, there they go!"

Why Leonard had broken with convention he'll never know, but now the keys were disappearing towards Calgary. He reeled around and dashed back to the road, waving frantically.

"Luckily Mary wasn't far down yet and luckily she looked in her rear-view mirror," he grimaces. "She turned around and came back and I retrieved the keys. It could've been embarrassing to say the least, with my truck right in front of the elevator doors and the keys on the way to Calgary."

As it was, Leonard and his intended provided a little entertainment for the farmers during their long and otherwise dull wait that morning.

"Norman and I did have the jump, being the first in line," says Leonard. "We got our trucks in and then we were on our way for two more loads. We hauled from the land closest to town, so we could make more trips. We hauled about three-quarters of the day before the elevator was plugged again. With twenty-six trucks in line when it opened, it didn't take long for the elevator to fill."

The Wolds also grew flax, an incredibly slippery grain. One year they hauled their flax in a less modern truck. The slide on the back of the wooden box wasn't wide enough for all the flax to pour out, so the men at the elevator had to climb into the box to finish the job with a broom.

"They sometimes used a ladder against the back of the box," says Leonard. "A new man who had come to work at the Alberta Wheat Pool put the ladder up and climbed up to clean out the corners of the box."

If he'd had any idea how slippery the grain was, he never would've rested the ladder on the loose flax spilled on the floor. Like tiny ball bearings, the flax whisked it out from beneath him before he had time to blink. In a split second the man was sprawled on the driveway floor.

"Fortunately only his pride was hurt," chuckles Leonard.

There wasn't much amusement the day a bin of flax exploded on Leonard's farm. The bin was wooden, bound with metal bands, and had been purchased from the Wheat Pool, where it was used to keep bulk fertilizer. The Wolds bought it to store flax. As storm clouds gathered menacingly on the horizon, Leonard and his youngest son Clarence worked to auger the flax into the bin.

"It was almost full and Clarence was at the top, watching so he would know when to shut the auger off," says Leonard. "Suddenly the bin exploded. It sounded like a cannon shot. Instantly Clarence was on the ground in the rubble of the bin and the pile of flax."

Clarence was all right, but the immediate question was why had the bin exploded?

"On investigation, we found that the bands on the inside had been corroded by the fertilizer."

The men were left staring at a terrible mess.

"When the bin burst, the flax spread like water over the ground."

The storm, brewing ever closer, made things worse.

"Water plus flax makes a really gooey mess," points out Leonard. "We feverishly began loading the flax from the ground into our trucks using a grain vac. Luckily the weather held."

But they had another problem.

"Since the bins were on a gravel base, it was impossible to keep all the gravel out."

They ended up with a bin full of salvaged flax seeded through with pebbles and stones.

"Our elevator agent, Gordy Lewis, was very co-operative," says Leonard. "In the winter during a quiet period, we hauled the flax into the elevator and very slowly and carefully ran it over screens to take out the gravel. We saved most of it, but it was a lot of work."

Leonard ponders how such a mishap might be handled today.

"We have always enjoyed the close personal relationship with our local country elevator agent. We often wonder how a large, high throughput elevator would react under similar circumstances."

# CHAPTER TWENTY-FOUR

# Swivelling

For little Kathryn Jessen, the lineups during busy times at the Sanford, Manitoba, elevator were interminable. The grain-filled trucks seemed to string out as far as her eyes could see. Small groups of men stood beside them on the dusty road, scuffing up gravel and shooting the breeze while they waited for the line to move. Kathryn was right in the middle of them, with her Mom or her Dad and their own grain truck.

"Often, it was my mother driving," she says. "She would stand and talk away to the other farmers just as my Dad did."

Her thick Scottish accent added interest to the talk of weather and grain prices, but Kathryn paid no attention. Her mind was on the sealer jar of homemade iced tea resting on the seat of the truck.

"It was a nice wet drink," she says, "but was never really even cool, because my mother didn't believe that ice cold drinks on a warm day were good for you. In Britain they never had ice in their drinks, so neither did we. All of a sudden there was a flurry of activity and we all hopped into the trucks, started them up, and moved one truck space forward. The engines were shut off, everyone

Sanford, MB.
PHOTOGRAPHER: UNKNOWN / CONTRIBUTOR: KATHRYN BARTMANOVICH

climbed out, and the visit resumed until it was time to move again."

When the Jessens finally got to the head of the line, Kathryn was fascinated by the unloading process.

"I stood mesmerized, watching the golden grain drop from the back of the truck into that mysterious place below."

She looked up at her parents, the agent, and everyone else who was there.

"The trains doesn't stop here anymore."
**WATERCOLOUR: KATHRYN BARTMANOVICH**

"They were covered with dust, dirt, and sweat. I remember thinking, 'How dirty Mom and Dad look.' Obviously I hadn't looked in the mirror."

Dust and sweat aside, Kathryn loved the farm.

"The best part of the day in the fall was after school and going to the field. I got to ride around and around the field on my Dad's knee. Sometimes he even let me steer, or so I thought. There were days Dad came home from town and said, 'There's a quota on.' I didn't understand what this meant. All I knew was there was a flurry of activity around the yard."

During the idle winter months, Kathryn loved being allowed to visit the elevator with her dad. She headed straight for her favourite chair in the corner of the office. Using her feet to propel herself faster and faster, she spun deliriously on its swivel base, listening to the men as their images blurred past her in a stream of colour.

When she got older, Kathryn was thrilled to be given responsibilities on the farm.

"I got to actually combine by myself," she says, "as Dad ate supper or it was time to move to a different field."

At long last she learned what a quota was: a special order from the Canadian Wheat Board allowing farmers to haul a certain type and amount of grain into the elevators.

Today the Sanford elevator is no longer open for business. Kathryn Bartmanovich is a grown woman, but the prairie way of life is still immensely important to her. She honours it in beautiful watercolour work.

"I am particularly fond of painting elevators, prairie scenes, and flowers," she says. "For me to see a grain elevator in the distance gives me a feeling of home wherever I am. I try to snap photos on my travels, as you never know if the next time you're through the area the elevator will still be there."

# CHAPTER TWENTY-FIVE

# Snakes in the Grass

L.P. (Bud) Hogue was no slouch. He started working for Pioneer Grain at a tiny siding near Sanctuary, Saskatchewan, in 1957. Thirty-five years later he retired from a position in the company's head office.

"I must say that I had many rewarding experiences in the grain trade," he relates. It is the not-so-rewarding ones he remembers best, though.

One example is the day the scale broke. A farmer with a full load of grain pulled into the elevator in the morning. He had just driven onto the scale when one of its corner bolts gave way. With a great crash, the whole works fell into the pit: farmer, truck, and grain.

"Needless to say, I was out of business for a few days," says Bud.

There was another time that he was involuntarily shut down. Businesses were supposed to remain closed on Sundays, but there is no such thing as a break during harvest. Bud was busy at work when he got some unexpected company.

"The RCMP came along and told me to close the doors, as I was breaking the law. All I was trying to do was help the farmers move some grain, as we had a very wet fall."

Bud obediently closed the doors. But he wasn't one to put up with nonsense.

"After the police left, I went back to buying grain," he states.

Not all of Bud's customers were co-operative. The odd shyster did his best to pull the wool over Bud's eyes. One farmer in partic-ular always seemed to have a lot of sand mixed in with his grain. Bud decided to get to the bottom of it. One day, when the fellow was hauling several loads, Bud went up to the top of the elevator and watched him from the cupola window.

"As he was passing the sand pit, he stopped and put two or three shovelfuls into his grain to offset the dockage that was being taken," says Bud. "This farmer also had a big dog that always stayed in the cab of the loaded truck until the truck was weighed. Then the dog was let out and not put back in until after the truck was emptied and weighed again."

This way, the farmer thought he was getting paid for the weight of the dog.

"Little did he know that we always compensated for his cunning ways," smiles Bud.

In 1972 Bud was promoted to the position of district manager. He now had more authority and miles of territory to cover. Authority brought with it difficult decisions. One in particular might have cost Bud his life.

"An elevator manager had a serious drinking problem and had his fingers in the till," he says.

Bud had no choice but to fire him. The man didn't just lose his job, he lost his home as well.

"When he had to move out of the company house, he was very upset," recalls Bud. In fact, he produced a gun and threatened to use it on the district manager! Fortunately, cooler heads prevailed.

"About one year later he joined Alcoholics Anonymous," Bud says. "He was hired by the government and he thanked me for terminating his position with the company."

In 1984 Bud was promoted again, this time to the Pioneer Grain Company's head office in Winnipeg, Manitoba. He spent eight years there before retiring in 1992.

Bud reminisces about the life of an elevator agent:

"We all suffered with skunks in the boot and with plugged-up legs and farmers all rushing in to have their grain tested in the fall," he says. "In Southern Alberta I contended with rattlesnakes in the grass at the back of an elevator while loading boxcars. They were left alone and disappeared when the weather turned cooler."

They weren't the only things that disappeared.

"I will say," comments Bud, "that I miss the many elevators that were landmarks for all of our small towns and communities."

# CHAPTER TWENTY-SIX

# Coffee, Oil, and Honey

"It took us all of twenty minutes to decide this was our ticket back to rural living," says Val Hvidston of the day her husband, Terry, was offered a job over the phone with United Grain Growers. The couple was living in Edmonton, Alberta. Both had "good jobs with promising futures," but a nagging conviction hounded them: neither wanted to raise a family in the city.

"We mentioned this to my dad, who lived in Saskatchewan," says Val. "One day he was having coffee with the superintendent from the UGG, who told him that they were having trouble finding grain buyers willing to move to small-town Saskatchewan. That night at 6 PM Terry received a phone call with a job offer."

Neither Terry nor UGG knew what they were getting, but UGG, at least, considered it a good deal.

"Terry's dad had been an agent for the Saskatchewan Wheat Pool for twenty-five years and Terry had spent many hours helping him. Terry's uncles had worked on the Wheat Pool construction crew."

Val too came from a rural background. She was stalwart, which was fortunate, for she was soon to discover the characteristic was

143

essential to being an elevator agent's wife. In May 1976 Val, Terry, and their newborn son set out from Edmonton for McMorran, Saskatchewan.

"We had a Sask Wheat Pool map," says Val, "and McMorran was on it, so we assumed it must be a normal little town."

How wrong they were!

"We followed the moving truck carrying our furniture to a little siding that had one house, one tree, and one elevator."

Of course the house was theirs, or more accurately, the company house they would live in while they were there. Val's heart tumbled to her toes when she saw the "little storey-and-a-half wartime-style house."

"The steps leaned one way and the door, which was missing the top hinge, leaned the other," she relates. "There was no paint on the house and no grass around the house."

Once they got over the initial shock, Val and Terry were undeterred. They dove straight into making the house not only habitable, but homey. They had a little trouble pulling off the habitable part.

"The sewer regularly backed up into the bathtub," recalls Val, "so the baby was bathed in a pan on the table."

But where there's love there's home, and the little family was soon snug and content. As Val made the transition from urban career woman to rural stay-at-home wife and mother, her staunch disposition came to the fore. In her lone house on the siding, she was almost completely isolated.

"Terry trained in Eston, twelve miles away, and we only had one vehicle, which he needed to get to work," she says. "I didn't know anyone and had no way of getting out and meeting anyone. So I absorbed myself in making our house a home and looking after Baby."

It is a fact that in rural Canada, even though not a dwelling may be seen for miles around, a vibrant social network still exists. Val was about to experience it.

"I had read in a magazine about how to make your hair shiny: steep honey and olive oil (half and half), massage it into your hair, wrap your head in Saran Wrap, and leave it for one hour. Then shampoo it out. I decided to try it. Wouldn't you know that was the afternoon that some of the neighbour ladies decided to drop in and welcome me to the community."

Val was mortified when, turbaned in Saran, she encountered them unannounced at the door. Her mind raced.

"I thought it would take too long to excuse myself and shampoo out the oil and honey because it would take several washings, so I made coffee and sat with a washcloth, wiping the drips that continually rolled down my face. I'm sure those ladies are still telling that story!"

Val didn't have long to worry about the impression she'd made. Her father had helped them with repairs to the house and he knew the conditions in which they lived. Without telling them, he contacted the UGG superintendent and complained.

"The superintendent promised Dad we wouldn't be in the house by winter," says Val. "True to his word, he offered Terry his own point by the fall of that year."

Mantario was just a short distance away but it was a real town.

"When our youngest son was born in 1977," says Val, "the population increased from thirty-one to thirty-two."

The Hvidstons adapted to their new surroundings, and Terry was soon busy at work in his new elevator. The Saskatchewan Wheat Pool was his competition but also his ally. The two agents collaborated whenever it was to their mutual benefit.

One thing they attempted was eradicating "the huge population of garter snakes that roamed between the elevators and the highway."

The Hvidstons on the steps of their UGG company house. The elevator across the street is reflected in the window. Mantario, SK
PHOTOGRAPHER: UNKNOWN. CONTRIBUTOR: VAL HVIDSTON

"The grass literally moved with them," says Val. "The men made a tool from a broom handle with a V-shaped, sharp metal end. As they killed the snakes, they flung them into a pile."

Little did they realize what would happen as the pile began to decompose. The stink was unbearable. The two had no choice but to deal with it. They must finish what they started.

"They had to shovel the snakes onto a truck and take them to the dump," says Val.

Terry bought grain in Mantario for almost two years before transferring to Beechy, where they could be nearer to family.

"Eventually Terry realized that he didn't want to buy grain forever and started taking university extension classes by correspondence."

Despite what Val describes as "lots of pencil throwing at midnight," he persevered and today is a rural municipal administrator in Tisdale, Saskatchewan.

"So we have maintained the rural way of life," she says. "But all of the elevators Terry worked at are demolished. When they were scheduled for demolition in Tisdale last year, we took some pictures—still feeling somewhat connected."

# CHAPTER TWENTY-SEVEN

# Career Aspirations

Kevin Marken eyed the 19 thousand dollars worth of Glean just delivered to the granary that served as his Pioneer elevator farm supply warehouse. That was a lot of money in the late 1980s and the lock on the bin was mighty flimsy by comparison. Everyone knew that farm chemicals were favourite targets for small-town thieves, and Schuler, Alberta, was no exception. Valued at four hundred dollars per one-litre jug, Glean was a particularly hot commodity.

Kevin's expression was thoughtful. He looked from the Glean to the 2-4D, a much cheaper product. Then he carefully unpacked both from their shipping cases. He hid the Glean in the 2-4D boxes, and pulled the same switch with other chemicals in the warehouse.

"A lock only keeps the honest people out," he says. "That very night burglars broke in. They ended up stealing mid-priced chemicals and bypassed the 2-4D Amine boxes where I hid the products that were more valuable."

Demonstrating that kind of thinking, it's no wonder Kevin had been promoted to manager after only three years as an elevator helper. When he took the position at Schuler in 1986, he was twenty-three years old, the youngest manager working for Pioneer Grain at that time.

Kevin likes to see silver linings. Even though the theft was devastating, he found great pleasure in thinking about those thieves.

"I bet they didn't have any herbicide receipts to use for deductions on their income tax returns for a few years," he smirks, "since they stole so much from my elevator as well as the other company in town."

Obviously the police weren't around that night. Usually they were conscientious about patrolling the elevators during chemical season. Leonard Wold can attest to that. He thought he was going to be hauled in and booked the night he decided to kill two birds with one stone.

"I had a half-ton load of empty spray jugs," he says. "I was at the neighbour's until after midnight and decided to come back through Alliance and unload them in the big container tank by the elevator."

Leonard's mind was still with his friends. He was completely absorbed in thoughts of their conversation as he absently tossed jugs into the tank.

"Suddenly light shone on me," says Leonard. He turned and froze, like a trapped deer. "It was two RCMP officers."

They coolly took in the situation, and fortunately, "right away saw what I was doing. They explained that this was the time of night some helped themselves to full jugs, rather than disposing of empties. The container tank was close to the shed where spray was kept, and I guess there'd been some spray stolen."

The Mounties immediately saw the humour in Leonard's predicament and broke into laughter. Leonard laughed too, even though he was still shaking from the scare.

There was no laughter at Kevin Marken's elevator the year it snowed in August. The crops around Schuler were tall and golden and harvest had just begun.

"I went after work to help a farmer who was a friend and customer with his swathing," he recalls.

Midsummer in southern Alberta is usually dry and hot, but this night the sky was grey and the wind blew bitterly cold. The men were glad to stop for the hot supper the farmer's wife brought to the field. Normally they would've eaten sitting against the machinery or on the open tailgate of a truck, but tonight that was out of the question.

"We set the lawn chairs up in the back of the half-ton," Kevin says. "It had a topper on it which kept the wind from blowing through our light coats."

They ate quickly and got straight back to work.

"We barely made it to the next corner before it started to rain. The belts were slipping and the crop wasn't cutting well, so we had to quit."

Kevin drove home in driving rain. He expected it to taper soon. After all, it was August. But instead the rain settled in and at some point during the night turned to snow. Kevin was stunned when he looked out his window the next morning.

"The heavy wet snow and the strong wind had laid down all of the crop—all in the same direction," he says. "Farmers scrambled to source a special swather attachment to pick it up."

Even so, the sodden crop, flattened to the ground, quickly lost value.

"Up until now I had generally always bought the grain for Number One in the Schuler area," says Kevin, "but because of the frost, green kernels, and stones, it wasn't top grade. It was hard for the farmers to take. The price of grain was low enough as it was.

This elevator stood abandoned for eight years. Salvador, SK
PHOTOGRAPHER/CONTRIBUTOR: CHRIS STACKHOUSE

This was the start of a very tough year in the elevator for me."

Nevertheless, Kevin gained a sense of comfort from waking in the night and hearing the shunting and bumping of the train spotting cars.

"It was always a good sound, especially when the elevator was full."

It meant, somehow, that all was right with the world. Occasionally, of course, he arrived at the elevator the next morning to discover all was not right.

"One of the more frustrating situations was to find that the train crew had spotted the cars on the wrong side of the loading spout."

This was cause for concern. The track was sloped slightly so that full cars could be rolled more easily away from the elevator, making room for the empty car next in line to take its place. Naturally, in order for this to work, the cars had to be placed on the uphill side of the loading spout.

"At most elevators," says Kevin, "there was a large arrow painted on the back wall above the railway tracks. This was to tell the train crew which way the cars rolled after they were loaded."

If the train inadvertently spotted cars to the right of the arrow, they had to be moved back up the slope to the left of it before they could be loaded and pinch jacked down the track. It was almost impossible to move a string of cars back up the grade manually. By that time most elevators had electric car pullers. Agents who didn't had big problems. Sometimes, according to Kevin, "the railway had to come back to spot the cars on the correct side before they could be loaded."

"One would think that a full grain car would easily roll over a thin layer of frost on the rails on its way down the gradual slope," he says. "Not so!"

All it took was a slight skiff or even just a frost to stop the cars in their tracks.

"The frost would melt just in front of the wheel from the pressure of the 260 thousand-pound car. It would then build up a ridge of ice and the car just wouldn't move."

Kevin learned the hard way that letting the brake off wasn't the solution. When he did succeed in setting the car rolling, the frosty rails took over and the car wouldn't stop!

"It often rolled past where you wanted it to. Then out would come the car jack to move it back again."

There was only one way to deal with the frost and snow. Get out the industrial-sized broom and sweep the tracks, just as though they were a sidewalk leading up the street.

By the late 1980s, grain companies were installing computers in every elevator throughout the prairies. There was no turning back. Agents who'd successfully bought grain for years with little more than paper, pencil, and scale stared at the high-tech equipment, so eerily human and utterly alien to their dirt-and-grain-strewn offices. Many were overwhelmed by the thought of having to learn a new way of doing things, complete with a whole new vocabulary. They felt totally defeated. Some retired early or took themselves out of management to avoid the computers. Others did their best to tackle the challenge, but never really caught on completely. They learned only what was needed in order to keep doing their jobs. There was no question that a new way of life in the elevators was emerging.

A few welcomed the change.

"I was always fascinated by these machines," says Kevin. "Even if they were slow. I once asked one of our company's computer support staff if the computer had enough space on it for me to use it for my personal accounting. He said that if I could figure out how to get my accounting on to this machine, he suspected that I wouldn't have to work in a dusty, hot elevator any more. He said that I could earn a lot more money programming the computer than just using it to buy grain!"

Kevin pondered these words over the next few years as he continued in the elevator. In 1994 he was just over thirty and the time for change had come. He left the grain industry to enter the burgeoning field of computers.

"My job is quite different now," he remarks. "I work in conditions that are cool in the summer and warm in the winter rather than the opposite, which often occurred in the old elevator office."

# CHAPTER TWENTY-EIGHT

# The First and the Last

"In 1924, the year I was born, my father Caspar became a founding member of the Saskatchewan Wheat Pool," says Walter Zunti of Salvador, Saskatchewan. Thus began a long-standing family loyalty to the co-op, broken only in the late 1940s when a new elevator became necessary.

"The old elevator was torn down and farmers could salvage the lumber for five dollars a load," recalls Walter. "In the fall of 1949, I hauled grain from the combine directly to Salvador with a team of horses. The new Pool wasn't quite ready to receive grain, so I hauled to the Alberta Pacific." The agent there had celebrity status. He was the brother of famous country and western singer Wilf Carter. But the Zuntis were not star-struck. They were diehard Pool fans. Came the day that Walter drove his team and wagon right past Alberta Pacific and on to the Pool.

Cliff Carter watched in puzzlement. The new elevator wasn't open yet. What was going on?

Walter drove his horses straight up and into the building where the agent, Ralph Doig, was waiting for him. A short time later, he drove the empty wagon out. He was the very first person to haul grain to the new Salvador Saskatchewan Wheat Pool elevator. The next day Ralph opened the doors for business, having with Walter's co-operation assured himself first that the scales were balanced and the equipment was working properly.

That winter, each with a team and bobsleigh, Walter and his brother hauled grain cross-country from their farm eight miles away. It was a long trip, broken only by the visit to the elevator and a meal at the local restaurant.

"The next year," says Walter, "the horses were retired and my father bought a Chev ton truck."

"Which I still use," he adds.

In 1986, when at thirty-seven the Salvador Wheat Pool was feeling its age, its sister elevator in Luseland, ten miles away, burned down. It seemed natural to replace it with the best facility money could buy. A contractor was hired and the first cement elevator in the province was erected.

"The grand opening was July 29, 1987," says Walter, "and the pressure was on to close Salvador."

Walter kept hauling there as long as he could. One day the hoist on his three-ton truck malfunctioned. He knew he could fix it, but in the meantime it wasn't a problem. He'd simply get the agent in Salvador to dump his load using the elevator's air hoist. When Walter got there, however, he discovered that the elevator was full and he'd have to haul to Luseland. Much to his disappointment, the new elevator turned out not to have a hoist. In that fine modern facility, Walter's entire load had to be shovelled off by hand.

"So much for technology," he grunts.

Not long after that the fateful day arrived.

"On July 31, 1992," says Walter, "the Salvador Wheat Pool committee organized a closing ceremony with a barbecue at the

The last load. Salvador, SK, 1992
PHOTOGRAPHER: UNKNOWN / CONTRIBUTOR: WALTER ZUNTI

Salvador Hall. Former grain buyers were invited. At 6 PM I drove the old Chev truck to the elevator with one hundred bushels of wheat. They used the air hoist to dump it, even though I had a hoist on the truck. The elevator was declared closed."

Walter had hauled its first and its last load. Though abandoned, the elevator stood for eight years, a comfortable and familiar anchor for all who gazed upon it. Then one overcast day in the fall of 2000, Walter looked up from summerfallowing and he couldn't see the elevator on the skyline anymore. Perhaps it was just obscured by the haze. The next day he checked again, but sure enough, it was gone. An acquaintance had loaded his camera with film to record the sad event.

"At 6 PM he went in for supper," says Walter. "At 7 PM the elevator was down—and he had no pictures! Salvador is now basically a ghost town."

# Epilogue
## Our Story

On a sleepy January afternoon in 1982, patrons stepping out of the tavern in Granum, Alberta, had reason to question their sobriety. They looked, blinked, then looked again as the roof of the Alberta Wheat Pool elevator across the street lifted several meters straight into the air. Then it settled silently down again. If not for the giant puff of brown dust that whoofed from its seams as it landed, they never would've known anything had happened at all. They looked at each other.

"Did you see that?"

Several minutes later I snapped out of a doze when Dale crashed through the door of our apartment.

"The elevator blew up!"

The Granum elevator fire of 1982 made news that night and for several days afterward. By the time the smoke had cleared, plans were underway to replace it by moving an existing elevator from Stavely, just up the track. That elevator was upgraded to modern standards, and the Wheat Pool was in business again. Within two years, however, the hotel from which the ogle-eyed patrons witnessed the first blaze caught fire itself. The fire jumped the street and lit up the new elevator. Once again it burned to the ground. The Alberta Wheat Pool admitted defeat and never built in Granum again.

By this time our little family was long gone. Within weeks of the first fire, Dale finished his training and was sent to Tofield as an assistant manager.

We were very excited to move into our first real home, an older one-and-a-half-storey company house, much like the one Val Hvidston describes. The partially finished basement had been taken over by some kind of fungi, and there were six-foot potato eyes growing out of the root cellar door. I never did muster the courage to peek into it. Within the first several weeks of living there, we fumigated twice for spiders, my greatest fear. But we loved being there.

To this home we brought Jiggs, a tiny, shivering puppy. He was black, with short legs, a long body, and brown, question-mark eyebrows. We got him for Brandi when she was two. It didn't take long, however, for him to discover his true calling as a diehard elevator dog.

Under the friendly and watchful eye of Stan Voltner, the station agent, Dale developed his skills in elevator management. Shortly after two years had gone by, Dale's regional manager approached him.

"We know you'd like an elevator of your own," he said. "We think you're ready. There's an opening in Gwynne. Would you like it?"

Would he? Dale jumped at the promotion. It didn't matter what the elevator or the company house looked like. We began packing, and in the summer of 1984 moved sight unseen to our new home.

Gwynne was a picturesque, peaceful hamlet of about seventy people, tucked into a valley between Camrose and Wetaskiwin. A school and a sprinkling of houses adorned the top of the hill, but in the valley below was the core of the village: two streets of houses separated by a large playground. We moved into the Alberta Wheat Pool house across from the playground and just up the street from Gwynne's single elevator.

The 1950s-style bungalow had a slightly better basement than the one we left in Tofield. An impressive feature was the massive cistern in one corner. I climbed onto a stool and peeked into it. Thick, reddish foam hovered over brown water. I scrambled down and never looked again. This was our tap water, pumped from a well in the backyard. Dale was instructed to treat it by periodically dumping a scoopful of black powder into the cistern. The label on the powder warned of possible death if the particles were inhaled or ingested in any way. I had the strange feeling we were poisoning ourselves. Every twenty-four hours the system churned to life, chasing murky water down pipes and through filters. We often woke at 3 AM to the sounds of this so-called cleansing.

We had no choice but to use the treated well water for bathing and doing dishes, but it didn't take us long to decide to haul our drinking and cooking water from Wetaskiwin, seven miles away.

I'd always done a lot of baking, but while in Gwynne I virtually stopped. The cupboards had bugs in them, and we couldn't afford fumigation. I couldn't stomach the thought of incorporating the critters into cookies, breads, and cakes. I did, however, make an effort for the Christmas cookie exchange that the postmistress hosted.

The social life in Gwynne was phenomenal. We were invited everywhere and fed wherever we went. The ladies met in each other's homes at least once a month, supposedly to celebrate birthdays. That was just the excuse; they would've got together anyway. Dale loved spending time at the curling rink, and he met several farmers who became good friends.

There was no store, gas station, or bank in Gwynne, but we

hardly noticed. Warmth of community spirit made up for the lack of services. The post office was an old portable trailer. It sat in the corner of the postmistress's grassy yard, across the street and beside the playground from our house. It had no running water, but that was no obstacle. She and I took turns carrying freshly brewed pots of tea over for delightful morning visits.

The pastor's wife next door to us was a treasure. A perceptive woman of few words, she seemed to anticipate my needs. She insisted on looking after Brandi when I came down with the flu, and she taught me how to make homemade chocolates. (In her kitchen, not mine.)

Other women too became good friends. They drew me in as though we'd known each other forever.

The winter was cruel. Our house was so cold that frost formed inside on the walls. I spent a great deal of time warming up in the bathtub, the very tub into which half the tub-surround tiles fell with a shocking crash in the middle of one night.

But Brandi learned to ski on the Gwynne ski hill, and there was a perfect slope for sledding right behind our house. In the mornings I stepped outside just to inhale the fragrant wood smoke that laced the air. Like Val and Terry, we were extremely happy despite the living arrangements.

Jiggs entered his teens in Gwynne and developed some very nasty habits. He ran the town looking for lady friends, and brought home treasures, such as other children's toys, the hindquarters of a dead beaver and a bra from the postmistress's clothesline. His chief delight was running cats up the nearest pole. I was quite relieved when he followed Dale to work one day and discovered his true calling. For the next twelve years he rarely missed a day at the elevator.

The Gwynne elevator, one of the oldest in the province, sat in the lowest part of the valley. It was never without water in the boot.

One of the last trains to enter the village. On their way to work
each day, Dale and Jiggs disappeared through the gap in the bushes
across from the house. Paradise Valley, AB, 1987
PHOTOGRAPHER/CONTRIBUTOR: ELIZABETH MCLACHLAN

A pump took care of most of it in the spring, summer, and fall. In
the winter, it wasn't a problem as the ground froze. The perpetual
damp, however, created a perfect environment for lizards. They
teemed underneath the pit. Out of sight, out of mind might have
worked nicely except that elevator agents must frequently crawl
beneath the elevator to check for leaks in the pit or accumulations
of grain around the scales. Fortunately Dale wasn't queasy. He
ignored them and they ignored him.

Dale enjoyed his work and was anxious to do well in his first
position as a manager. The farmers were quite willing to let him. The
elevator phone was connected to the house—and they knew it. It
was never a problem, as far as they were concerned, to call at 11 PM
or any other hour.

"I just have one more load to haul," they'd say.

Gwynne's elevator days were numbered. Soon, the point would be designated a satellite to Wetaskiwin, then closed, then demolished.

Dale's career, however, was just beginning. In the spring of 1985 Brandi and I left for three weeks in England. By the time we returned, Dale's application for Paradise Valley had been accepted. We had one week to move. In a station wagon loaded down with the last remnants of our existence in Gwynne, I cried as I drove away from one of the best periods of my life.

Two things I learned well as an elevator agent's wife: one week is ample time to make a major move, and regardless of what you leave behind, adventure and friendship always lie ahead.

As its name suggests, Paradise Valley was also picturesquely nestled in a valley. We had graduated, however, from seventy people to two hundred, and from one old elevator to four. The house was a little newer and a little better. Every day Brandi and I watched Jiggs and Dale disappear through a gap in the bushes across the road. From there, they followed a narrow path along a small slough and up to the tracks. A short walk down the tracks brought them to the elevators.

By 1985 tanker cars had almost universally replaced boxcars. With roughly 40 per cent more capacity, they were round, with funnel-shaped, hoppered bottoms. Best of all, they didn't need coopering. Dale had coopered only a few cars since his career began, but that was about to change. Paradise Valley was on a branch line with two other points, McLaughlin and Rivercourse. The track was light steel that couldn't handle big cars. So the only kind the train brought in were boxcars. Back to coopering!

The large, cardboard grain doors were in demand for other things as well. Schools loved them for creating scenery and backdrops for plays and special events. They often called looking for donations, and

Dale was happy to oblige. Dale's helper, Bill, also took advantage of the clean, crisp cardboard, but in a more unusual way.

Like Tofield and Gwynne before them, the Paradise Valley elevators had no running water. An outhouse fitted with a bucket had to do. The Paradise Valley outhouse was a disgrace and hadn't been used for a long time. At age four, Brandi was becoming familiar with what she referred to as her dad's big green alligator. She loved to spend time there with him in the evenings. The thrill of racing up the manlift encircled securely in his arms was glorious. She was terrified, however, of the outhouse. According to her, it was all "dust and spiders and snakes crawling up the hole."

One day she got a lovely surprise. Bill, a talented artist, had transformed the biffy into a real house! He lined the inside with coopering doors, then used them as a canvas to paint a cheery room,

Minutes after the last train leaves Paradise Valley, CP Rail employees inspect the track in a jigger, in preparation for tearing it out. Paradise Valley, AB, 1987
PHOTOGRAPHER/CONTRIBUTOR: ELIZABETH MCLACHLAN

complete with curtained windows framing outdoor scenes. It was the prettiest outhouse in the province; in a flash, one little girl's fears were banished.

I doubt there were snakes in that outhouse, but other creatures lurked. Deer came to feed on the spilled grain behind the elevator, and a couple of beaver lived in the slough between our house and the railroad tracks. One day I encountered a mother cow bawling at me through the kitchen door. She was clearly distressed. I called Dale at the elevator. He investigated and found that the cow belonged to some neighbours up the road. She'd lost her calf and was searching mournfully for it. As a mother, I identified. Dale came home and slowly we herded her back to her grateful owners.

There wasn't much we could do, however, when the same neighbours accidentally let their ferrets loose and they took up residence under one of Dale's elevators. They dug in for weeks. Jiggs went below to make their acquaintance one day and was quickly sent packing. Eventually their owner was able to lure them out, despite the attraction of the dark and the abundance of mice to eat. Other tenants were a family of foxes. Perhaps Jiggs had learned his lesson. He stayed away from them.

One lesson Jiggs never did learn, however, was to steer clear of skunks. Every time he got sprayed he suffered the humiliation of several baths. The first one involved two litres of tomato juice. It was followed by a rinse into which a whole bottle of peppermint extract was dumped. Finally he got a proper shampooing with dog soap, and plenty of vigorous scrubbing. Perhaps for him the most humiliating aspect of the whole ordeal was how pretty he smelled afterwards.

Since Jiggs had abandoned Brandi for elevator life, we decided to get her a cat. We named the black kitten Gert because there were so many adjectives that applied—galloping, gobbling, and gorgeous, to name a few. Gert quickly grew up and gallivanted off for three

Climb Through Time museum. Note that the loading spout is
gone and so are the tracks. There is no more spilled grain
for the deer to come and feed on. Paradise Valley, AB
PHOTOGRAPHER/CONTRIBUTOR: CHRIS STACKHOUSE

days with the neighbouring Tom. Shortly thereafter, she presented
us with kittens, one of which was gargantuan. We named him Big
Black Bart. He and Jiggs became fast friends and wrestling partners.
Soon Black Bart was joining Jiggs and Dale on their daily treks to
and from the elevator. It was a sight to see them picking their way
along the tracks single file. First Dale, then Jiggs, then Bart.

Paradise Valley's days too were numbered. On a foggy bleak day
in November 1987, the train entered the village for the last time
before the line was closed. The tracks were ripped up and the elevators in Paradise Valley, McLaughlin, and Rivercourse torn down. All
but one. The citizens of Paradise Valley rallied together, determined
to save their last elevator. Nearly every one of them volunteered both
time and money. With the addition of grants, almost a quarter of a

million dollars was raised to convert the elevator into a beautiful museum and teahouse. Aptly named Climb Through Time, the restored facility now enthralls thousands of visitors each year.

Between Paradise Valley and McLaughlin, the Alberta Wheat Pool built a new inland trucking facility. Its name, Three Cities, commemorated the three towns along the defunct rail line. From there grain was trucked to Kitscoty and Lloydminster, both elevators on the main CNR line.

With the closure of Paradise Valley, Dale was asked to take the elevators at Islay. Sadly, Black Bart would not come with us. He'd been running across the road one day to play near the slough—and he never made it. We consoled ourselves with the knowledge that his brief life had been extremely happy.

Islay also had a population of two hundred, but its two elevators were in better shape than those in Paradise Valley and Gwynne. The company house, built in the early 1960s, was the best yet. Shortly after we moved in, the maintenance crew painted it and carpeted the bedrooms. We were really moving up in the world!

By now Brandi was in grade two. There was no school in Islay, so she took the bus to Kitscoty. I was taking university transfer courses in Lloydminster and eventually went to work for the Crisis Line in Vermilion. I wasn't always there when Brandi returned from school. In such cases, she had instructions to phone Dale as soon as she got home.

"Dale's Pool Hall. Who in the hall do you want?" Dale always answered.

Years later we discovered how mortified Brandi was by that greeting. But Dale had seen the bus come in and knew exactly who was calling. He told her to come straight to the elevator. She hopped on her bike and rode to the end of the street, then crossed the tracks and was there.

The thrill of hanging out at the elevator never waned. Often Dale put her to work on the adding machine or had her sweep the driveway with a giant broom. She was a little nervous of stepping on the grating over the pit. What if she fell through the cracks? It was titillatingly scary.

Brandi admired her father tremendously. The first time she saw him tie a rope onto a string of grain cars and crank them down the track, she thought he was superman. He also let her drive the 1964 Chevy he used for work. Perched in his lap, she proudly steered the truck up and down elevator row while he manned the iron rod that stuck out of the floor in lieu of a gas pedal. Jiggs ran along behind, a huge grin plastered across his face.

Jiggs was a great hunter. He sat stock-still for hours waiting for gophers to come out of their holes. If he saw his prey escape into one end of a culvert, he dashed to the other end and stood guard, even if it took all afternoon. He disappeared under the elevator for long stretches. One day he frightened the wits out of a crew member who was under there doing repairs. All the poor fellow saw was a ghostly shadow moving about in the semi-darkness. Dale had a job convincing him it was simply the dog. Or was it?

One day Dale and his helper were cleaning the boot. Irvin was filling the pails and Dale was pulling them. Suddenly Irvin came flying up the ladder.

"I just killed a rat!" he shouted.

Dale was stunned, not because rats were new to him, but because there weren't supposed to be any in Alberta.

"Are you sure?"

"Yes, I'm sure. I killed it with my shovel."

"You'd better show me," said Dale.

But Irvin refused to go back into the boot. Dale took the flashlight and went down himself. There was the dead rat. Dale peered beneath the pit. Tracks ran all over. He climbed back out of the boot and called the provincial rat patrol.

"There are rats in my elevator," he told the lady who answered.

"No there aren't," she said. "Alberta doesn't have rats."

"I tell you I have rats. My second man just killed one."

"It was probably a muskrat," said the lady. "Islay is beside a slough. It's a common mistake."

"Listen," said Dale. "I grew up in Saskatchewan. I know what a rat looks like and I tell you there are rats in this elevator."

Reluctantly she agreed to send someone out. The minute the man saw Irvin's rat, he knew he was dealing with the real thing. He donned a large stethoscope with a long copper stem. With doctor-like demeanour he walked around each elevator, applying it to the walls and listening intently. Then he went down into the boots and did the same. He declared Dale's second elevator clean, but the scrabbling and squeaking he heard within the walls of the main elevator left no doubt.

"I want to check underneath this elevator," he said to Dale.

Dale showed him the three-foot door on the side of the building.

"It's already open. My dog goes in there every day," he said.

Within minutes the man was back.

"You've got quite a dog," he said. "There's a whole slug of dead rats under this elevator. He's been killing them."

That same day the man placed feeding stations of poisoned water in the boot and under the elevator. They were partially hidden so as not to look exposed. Rats were intelligent creatures, he explained. If the stations were out in the open, they'd get suspicious and stay away from them. Likewise, if the first rat died before the last rat took the poison, the rest would catch on and not drink the water. Therefore the poison was slow-acting.

"Nobody goes near those stations for ten days," he ordered. "That includes the dog."

Exactly ten days later the rat patrol was back picking up dead rats of all shapes and sizes. There were at least three hundred, spanning three generations. So many that the rat man was certain they'd been preparing to split and head farther down the track to form additional communities elsewhere. Thank goodness Dale and Irvin grew up in Saskatchewan.

Having lost its school and several businesses, Islay was struggling. But it wasn't dead. It still had a post office, small general store, curling rink, and modern hospital. It didn't have a coffee shop, but the elevator office was the accepted substitute. Every day Dale brewed a thirty-cup urn that wheezed and moaned throughout the morning while the men of the town drifted in to catch up on the latest. If Dale and his helper were busy, the men made their own coffee. It was a great arrangement. The only problem was a sorry lack of privacy, especially if Dale's supervisor happened to appear.

"He knew not to come during coffee time," says Dale. "If he did, he knew he was going to sit around and have coffee with everyone in town."

The exception was bonspiel season. The people of Islay were crazy about curling, so the elevator office emptied as everyone beelined for the curling rink. Even Dale headed over, either leaving his helper to hold down the fort or simply hanging a sign on the door. Every farmer who wasn't already at the rink knew where to find him anyway.

Occasionally skipping out to curl in a bonspiel probably did Dale's business more good than harm. In 1990 he was one of only fifteen Alberta Wheat Pool managers who reached their grain handling goals. Islay also made the highest profit on Agro products of all the stations in the territory.

Dale loved Islay and made many good friends. Several of his farmers were avid hunters, and it pleased him to be invited along on their annual ten-day moose hunts in the Peace River District. It was the farthest north he'd ever been in Alberta.

But the winds of change were blowing through the Alberta Wheat Pool. Cutbacks and layoffs were rampant; stations were consolidated; elevators began to come down. The second elevator at Islay was one of them. A special crew came in and took two days to complete the job. It was my first experience seeing an elevator crumble and fall.

The remaining Islay elevator was scheduled to be satellited to the larger terminal ten miles away in Vermilion. What would become of Dale? We had learned by now that we had little control over our destiny with the company. The worst thing in the world is having to simply sit and wait while others decide your fate.

In the summer of 1991 we were vacationing at our cabin in Cochin, Saskatchewan, when our neighbour came to the door. Someone in Alberta was trying to reach Dale, and since we didn't have a phone at the lake, they'd tracked down the man next door, who did. Mystified, Dale returned the call. It wasn't his boss. It was his boss's boss. He wanted Dale up at the Alberta Wheat Pool offices in Grande Prairie to interview for a position as grain manager in Falher.

We were stunned as we quickly packed and closed the cabin for the winter. There was no way of knowing when, or if, we'd be back.

Brandi was not amused. Two and a half hours home to Islay, then nine hours up to Grande Prairie was a lot of driving. Dale had his interview, and we drove two hours out to Falher to look at the town. The company was getting out of providing employee housing, so we talked to a real estate agent, then spent the night in a quaint motel before driving back to Grande Prairie. Dale was offered the position. We immediately drove back to Falher and bought a house, then we returned to Islay. Barely two days had passed since the phone call.

By now the '64 Chevy had been replaced by a '70 Duster. Dale packed it with bachelor supplies, and Jiggs, and the two set out to begin work in Falher. Brandi and I stayed behind with Gert and her second offspring, Zeb, to wrap up home affairs. Three days later, we rolled sleeping bags out onto an empty living room floor and tried our best to get some sleep before striking out for Falher early the next morning.

The cats were suspicious.. Strange men had invaded the house and taken everything away. Something was up. I congratulated myself on having the foresight to talk to a vet. An hour before our departure, Brandi and I gave the cats tranquilizers.

What a mistake! Gert reacted badly and Zeb hardly reacted at all. He set up caterwauling as soon as we put him in the car. For seven hours he kept us informed of his displeasure. Gert, I think, was allergic to the pills. She morphed into a feline Jekyll and Hyde and spent the trip alternating between complete relaxation in Brandi's arms, and wild-eyed aggression towards everything in sight, teeth and claws fully bared. We didn't know from one second to the next what to expect. Why we didn't have pet carriers is a mystery to me now. Of course we couldn't let the cats out of the car. Despite our careful planning, Gert needed the litter box. She tried to dig a hole in the burgundy carpet of the Oldsmobile, and when that didn't work she had diarrhea on the floor. It added a new sensory dimension to the trip. I've never been so glad to see anything as I was to glimpse the Falher elevators on the horizon that day.

All six of us stayed several nights in the local motel. Frazzled, and sick of restaurant food, we finally took possession of our house on the first of September. It was pouring rain. On September 2, we delighted in Kraft Dinner for supper and completely forgot about our wedding anniversary.

Falher, a community of twelve hundred in the Peace River District of northern Alberta, was settled by French Canadians almost a hundred years ago. Most of its population can speak English, but their first language is French, or rather, an interesting combination of French and English described as franglais. We, and others like us, were immediately labelled anglophones. Any English spoken at any time was done so only in deference to us. Even then, there was a tendency to forget and slip into the native tongue. My high school French was sorely taxed the whole time we were there. Fortunately, we were welcomed with open arms and for four years treated to the incredible hospitality, humour, and at times hothead-edness those of French heritage are famous for.

One of our first experiences was the Alberta Wheat Pool Christmas party held in the home of one of the employees. I dreaded it. In the past, staff parties had entailed sitting for hours in a dimly lit, smoke-filled room, the alcohol flowing, the music blaring, and the men talking only about work. The faux New Year's Eve parties in one town Kevin Marken lived in must have been similar. July 31 was the end of the crop year, so in this town, all the elevator managers got together to celebrate.

"I thought this was so funny," Kevin says. "These men couldn't stand each other the rest of the year, yet they would try to have a barbecue together on what they called New Year's Eve. It started out quiet until someone brought up the subject of work and then the arguments were on!"

But how little I knew about Franco-Albertans. At our party, there was no music, the lights stayed on, and we were kept busy playing games all night.

As a recent addition to the staff, Dale was the evening's victim.

His co-workers blindfolded him, stripped him of one of his socks, and placed his bare foot in a basin of ice water. Then they set a wooden stump before him and put a carrot in one of his hands and an axe in the other.

"Feel the carrot carefully," they instructed, "and try to find the middle, because you have to chop it exactly in half on this chopping block."

Two assistants guided him so that no body parts accidentally went missing. Time and again Dale set the carrot on the stump, calculated where he thought the halfway point was, then raised the axe and brought it down. Each time they told him he hadn't quite hit the halfway mark and gave him another carrot to try again.

What he didn't know was that every time he raised the axe, the assistants whipped the carrot from the chopping block and replaced it with his sock! He chopped his sock to bits! I was appalled, but I had to admit it was funny. The fact that they cared nothing for a good pair of socks was my initiation to the often irreverent francophone humour.

At midnight, our host set the coffee in a conspicuous spot and put out plenty of food. It was a clear signal to sober up. We ended up staying until 4 AM, drinking coffee and talking. It was the beginning of fine friendships.

Twice more while we were in the Peace, the carrot and axe ruse was perpetrated upon unsuspecting souls. Once it was the president of the company at a regional Christmas party in Grande Prairie. It was disguised as an initiation into the esteemed order of the fictitious Northern Knights. Another time, the Falher district entered a team in an Alberta Wheat Pool hockey tournament in Stettler. They targeted one of the other players, but the joke was on them for once. When the poor man's foot was thrust into the ice water, he panicked. He jumped up, threw off the blindfold, and before the astonished eyes of his captors, lit out the sliding doors of the hotel room and jumped off the balcony.

Whether it was the unique disposition of the people, the isolation of a northern community, or both, we soon learned that nothing was ever done by halves. People worked, lived, loved, and played hard. The Alberta Wheat Pool staff in Falher and the surrounding communities of Girouxville, Donnelly, and Nampa put together both a hockey team and a ball team, as well as a golf team and a rink of curlers. Dale was involved with them all. It meant many late nights and many hours on the road. Eventually, we thought nothing of driving wherever we had to go at any time. Grande Prairie, one hundred twenty-five miles away, was just a hop, skip, and a jump. Road conditions, too, were hardly worth a mention. Long, dark winters with masses of snow and ice made them treacherous, but that didn't stop anyone from using them. Life merely carried on.

Brandi walked to school in the dark on winter mornings. There were two programs at the kindergarten to grade twelve Community School: French Immersion, and English with a compulsory French component. Since Falher was a Catholic community, religion was compulsory too. By the time Brandi walked home again at half past three, dusk was falling. One day, with a flash of insight, I suddenly understood why everyone in the region painted the interiors of their houses white.

Dale had come a long way from a single elevator with a bucket in the outhouse. Falher was a large, modern facility with a minimum staff of six. There were showers and a conference room in the basement, not that they were used much. They collected dust and dirt just like every other nook and cranny in every other elevator everywhere. Brandi gazed at the shower stall and tried to imagine the feasibility of taking a shower without a curtain.

Not long after we arrived, the assistant manager at the station moved away. Of course there was a farewell gathering. It tied in nicely with a mid-week supper meeting scheduled for the conference room. Dale came home at 5 PM, cleaned up, then headed back to the elevator at 6. It was nice to have the district meeting right in town. In the dead of winter he didn't have to bundle up and drive off to some distant place.

After the meeting, the merrymaking began. At midnight, when Dale wasn't home, I thought nothing of it. If he wanted to stay out until all hours on a work night it was his problem, not mine. When I awoke at 1 AM and he still wasn't home, I again thought nothing of it. At 2 AM the phone rang. It was the wife of one of the other staff members. Roger wasn't home yet and there was no answer at the elevator. She was in a panic. I was more annoyed. They were known to move the festivities over to a house and carry on indefinitely, but on a weeknight? How crazy can you get? By 3 AM, when neither husband had yet to materialize, I started to worry just a little. It was, after all, thirty degrees below zero.

Dale, in the meantime, was having his own adventure. The party was indeed going strong at midnight when Cliff, who was new to the region, got up to leave. Six miles out of town he hit the ditch, which in the Peace in the winter means ploughing into several feet of snow. Cliff left his car and walked to the nearest phone, where he called back to the elevator for assistance. Dale and Roger volunteered. They found Cliff and made swift work of pulling him out with the winch on Roger's Jeep. Then they saw him safely home. On their way back into town, however, they lost control and hit the ditch themselves. They weren't stuck, but the Jeep couldn't climb the icy slope back to the road. Roger's winch couldn't help them now. By this time it was 2 AM. The party at the elevator was long over. While Roger worked to free his Jeep, Dale walked in his running shoes and light jacket a mile and a half back to Cliff's. The roles were reversed as Cliff crawled out of bed and came to their rescue. He hooked a chain to the Jeep and pulled it from the ditch. It was 3:30 AM before Dale straggled in. His feet, along with the rest of him, were miraculously unscathed.

In the relative isolation of the north, help wasn't always close, but it could always be counted on. Falher was surrounded by acres of farmland, within twenty miles of which were thick forests and bush. Dale and Brandi often camped in the bush, frequently taking a borrowed truck and an all-terrain vehicle. One morning they woke to steady, heavy rain. Disappointed, they broke camp to head for home. The truck, however, wasn't going anywhere. Its wheels spun deeper and deeper into the mire, until they were hopelessly stuck. Undaunted, Dale unloaded the ATV. They'd four-wheel drive to the nearest farm to get help. Over and over he pushed the starter button. The machine wouldn't even sputter. Dale and Brandi set out walking in the pouring rain. It was four miles to the farm. A little more than three miles into the trek, they spotted a truck. It was sitting empty in one of the numerous fields carved into the bush, and Dale recognized it as belonging to one of his customers. Without a moment's hesitation, he packed Brandi into it, and finding the key in the ignition, drove the rest of the way. The farmer was no doubt shocked to see his truck coming up the driveway with someone else behind the wheel, but such things were always taken in stride. He gave the sodden duo a ride back to camp in his tractor, which he then used to pull their truck from the mud. With friendly waves, they all went on their way.

A sense of co-operation and camaraderie helped us all get through the long winters. Many times the mercury dropped below minus forty. Dale followed the example of the other men and grew a beard for warmth. One day, he went out to shoo the pigeons from the elevator driveway and found they wouldn't move. Their feet had frozen to their perches and they'd died standing up. Jiggs was in similar danger. Dale had to watch him closely when he went outside because his paws froze up and he simply halted in his tracks, unable to move until someone rescued him.

As harsh as the winters were, the summers made up for them. One evening in mid-May, Brandi and I joined Dale as he delivered seed to a farmer thirty miles away. Driving back to Falher around 11 PM, we marveled at the bright orange sun just setting on the horizon. By 3:30 AM it started to rise again. As spring ripened to summer, I took many midnight walks in broad daylight.

It must have been hard for the thieves to have such a small window of darkness in which to operate. Every spring they raided elevator warehouses for chemicals, which were then sold on a black market system. Falher, as the main warehouse for the region, had a sophisticated alarm system. Even so, Dale often left the office light on all night and patrolled the elevator after hours.

Late in May, the Alberta Wheat Pool entered a team in a local slow-pitch tournament. When he wasn't playing ball, Dale was umping, and when he wasn't umping, he was setting up and helping to man the concession booth. He was exhausted as he crawled into bed Saturday night.

But just as a mother listens for her baby in her sleep, Dale listened for his elevator. At 1 AM he was out of bed and at the phone almost before it rang.

"The alarm is going off in your warehouse," said the Security Company in Peace River. "Would you like us to call the police?"

"Definitely," said Dale. He hung up the phone and threw on some clothes. "I'll pick up Dennis on my way," he said as he raced out the door.

Dennis, Dale's assistant manager, lived only half a block away. The RCMP, on the other hand, came from the detachment in the small town of McLennan, twenty miles up the highway.

When Dale and Dennis reached the elevator, all was quiet. They kept a respectful distance from the warehouse while they waited for the police. Not three minutes later, all hell broke loose as a patrol car came screaming into the yard. The officer later told the boys he'd come from McLennan at one hundred thirty miles an hour, actually becoming airborne as he crossed the intersection of Highways 2 and 49. The car screeched to a halt, and the trunk flew open.

"I gotta get Betsy," announced the officer as he strode to the back of the cruiser and pulled out a shotgun, which he cocked in one swift motion.

"Are you ready, boys? We're going in."

Dale and Dennis had counted neither on a gun nor an over-charged police officer. They were far more terrified of Betsy than they were of the culprits. Quivering, they followed the gun and her officer to the warehouse. The door was locked, so they opened it with the key and stood back to let Betsy through. Inside, the small red lights of the alarm system were flashing but everything else was silent and still. They cautiously advanced, Betsy at the ready in front of her man, and Dale and Dennis behind him.

Suddenly the wind caught the open door and slammed it shut. The officer wheeled around and for one horrifying moment, Dale and Dennis found themselves the target of Betsy's gleaming eye. A split-second later they were flat on the floor. But Betsy wasn't looking for them. Having found the sound harmless, she turned back to the trail of the thieves. Dale and Dennis picked themselves up off the floor and reluctantly followed.

A minute later they found a makeshift ladder, then a discarded flashlight, and finally at the back of the warehouse, a bashed-in door. The officer stepped out of the gaping hole and shone his light over the tracks. Three sets of footprints sprinted off into the field.

Within minutes, the town was crawling with police cruisers.

They stopped everyone who was moving about and set up road-blocks at all the town entrances. Dale and Dennis were left alone at the warehouse to repair the damage and do a quick inventory.

It was the night's darkest hour. With a board slung over his shoulder, Dale trudged up to the warehouse. Right behind him came Dennis, carrying a skill saw in one hand and a hammer in the other. Their actions looked decidedly shady. They went into the warehouse and set to work on the door.

A police cruiser rolled silently into the yard.

Minutes later, when Dale and Dennis emerged, they found themselves staring down the barrel of yet another gun.

"Who are you guys and what are you doing?" barked the police-man.

Dale and Dennis scrambled to profess their innocence, but they got nowhere. They couldn't even produce ID. In their haste to get down to the elevator, they'd left their wallets behind. The officer grew more suspicious, and more hostile, by the minute.

Finally Dennis said, "I know the first officer who was here. Call him on your radio. He'll tell you who we are."

Keeping a close eye on his culprits, officer number two reached for his radio.

"One man is Dennis Simard," said Betsy's owner. "I don't know the other one, but he's bigger."

The officer shone his flashlight directly at Dennis.

"You!" he ordered.

"Dennis Simard," trembled Dennis.

Then he shone his light on Dale.

"And you must be the big one!"

That was good enough for him. In the clear once again, Dale and Dennis proceeded with their task. It was 4 AM by the time they finished, and because neither of them was going to sleep anyway,

This old elevator was gone by the time we reached Falher. Note, however, the repair crew trailer in the left foreground (far more luxurious than the one the boys converted into a hunting shack). Falher, AB

PHOTOGRAPHER: UNKNOWN / CONTRIBUTOR: ADDIE LEVY

they went to the all-night truck stop for breakfast. By 8 AM they were back at the elevator serving a farmer who wanted seed; by 10:30 they were at the slow-pitch tournament, playing ball once again.

Their encounter with the law, and near encounter with the lawless, may have frightened the boys that night, but I doubt it scared them as much as Dale scared me several weeks later. He wasn't due home for lunch for another two hours, but suddenly he was rushing through the door, looking like a creature from a horror movie. His face and hands were a raw, livid red. In one glance I knew he'd been splashed with liquid chemical. I was sickened by the implications. Chemicals in elevators were hazardous. They consisted of pesticides, fungicides, herbicides, and other noxious substances. If they weren't corrosive, they were toxic. Rigid rules were in place for handling them with the utmost caution. Dale quickly explained that he'd been working in the warehouse when a barrel of Vitavax burst, spraying him in the face, hands, and all down his front. I was sure he had severe chemical burns. Then he cracked a smile, strode into the bathroom, and washed it all off. It turned out that this particular chemical was relatively harmless. Instead of getting burns, he got stains. The bright-red liquid eventually washed out of his hair and skin, but never did leave his clothes.

I wonder what the deer thought when they saw a hunter with vivid red splashes all over his coveralls? Hunting was a major activity in the Peace. Almost everyone was the proud owner of a crude hunting shack somewhere out in the bush. They ranged from old trailers to lean-tos and consisted of the bare necessities: beds, along with a wood stove inside; a campfire, clothesline, and a biffy outside. Roger's was the exception. It was an old Alberta Wheat Pool chemical shed from Falher, fixed up like a regular trapper's cabin. But he had it tucked so far back in the bush that unless the river was frozen, it was only accessible by all-terrain vehicle. The boys at the elevator had access to Dennis's cousin's shack, an old travel trailer impressively camouflaged to blend with the trees, but they sorely desired a shack of their own. The opportunity came when the repair crew from High Level acquired a new ATCO trailer. Their old one was up for grabs if any Alberta Wheat Pool employees wanted it. The boys in Falher jumped at the chance. I was mortified when the decrepit, dirty gray hulk was delivered to the elevator. It was a wreck for which shack was far too kind a term. I immediately had visions of all the wives being recruited with buckets and mops. I made my stance loudly known: I would not set foot in that creepy tin can. That was okay. No feminine input was wanted anyway.

The guys were thrilled beyond words. With obvious relish, they

View from the golf course. On slow days, Dale went golfing. If a
customer arrived, he drove the golf cart straight off the fairway
and back to the elevator. Provost, AB
**PHOTOGRAPHER/CONTRIBUTOR: CHRIS STACKHOUSE**

set about cleaning, repairing, and painting. For a solid week they
devoted their evenings to it. The final touch was the installation of
old metal bunkbeds salvaged from another repair crew trailer. It was
Saturday. Moose-hunting season had just begun. They carted their
prize out to the special spot they'd picked for it and thus began what
turned out to be a very worthwhile venture.

The shack wasn't just used for hunting, and while I stuck to my
word and avoided it at all costs, Brandi and Dale spent many happy
hours there, bonding over the campfire on ATV expeditions and
overnight campouts.

Of course, Jiggs was always along. He was a steady companion
who, as he aged, grew quite complacent about his position as
beloved pet and permanent elevator fixture. One day a farmer, who
was also an AWP delegate with some clout, complained. It seems
Jiggs was taking the best chair in the office for his own, making him-

Dale's last elevator point before bowing to change and taking a severance package from the company. All of Coronation's elevators are now gone. Coronation, AB
PHOTOGRAPHER/CONTRIBUTOR: CHRIS STACKHOUSE

self comfortable in it for most of the day. It was true. Dale promptly expelled him from the chair and arranged a bed for him in the corner behind his desk. Jiggs's nose was out of joint. Not only had he been banished from his favourite chair, but he'd also been hidden out of sight. He didn't realize that he was lucky. The farmer didn't feel that the dog should be at the elevator at all. No doubt Roger, in sympathy, gave Jiggs two sugar cubes that day instead of the usual one he got at coffee time. It was small compensation.

After four years in Falher, it seemed time to move on. Dale applied for elevator manager in Provost. Once again, we staged a whirlwind move in one week. This time, however, we weren't so hasty about buying a house. Provost was a booming oil town and the market was tight to say the least. We boarded our cats with friends, and courtesy of the Alberta Wheat Pool, Dale, Brandi, Jiggs, and I stayed in a local motel for a month. No cooking, laundry, or housework for an entire month sounds like heaven, but it wore pretty thin. We were glad to finally find a house and take possession in July 1995.

Provost was the largest grain handling point in a region encompassing over twenty other stations. Its three elevators, with staff of seven, were situated on the southern edge of town. The greatest thing about the location was the golf course. The number one tee-off box was a mere fifty yards from the elevator office. Dale got into the habit of getting up at 6 AM and shooting a round of golf before work. If it was a slow day, he took the cellphone and left instructions for his secretary to call if anyone showed up. When she did, he drove the golf cart straight off the fairway and back to the elevator. In the evening, he sometimes shot a round before coming home.

In December 1996 we lost Jiggs to arthritis and old age. Nine months later he returned as a cat that wandered into the elevator one day. It was as if he'd just wakened from sunning himself in the driveway. The little orange Tom endeared himself to all the staff and attached himself to Dale. After trying on names such as Dick, Herbie, Scooter, and Kitty he landed on Minou, the term our franglais friends in Falher used for kitten. Minou's behaviour was almost identical to Jiggs's. He hunted birds and mice and stalked gophers with exactly the same technique. He followed Dale everywhere, just as Jiggs had done, and when Dale finally broke down and brought him home, he dashed out to the van every day, exactly at 5 PM, just as Jiggs had. He even slept, like Jiggs, on Dale's clothes beside the bed and in his favourite chair at the elevator.

If Dale happened to be in the chair, it was no problem. Minou simply climbed up behind Dale's back and snuggled himself in for naps. One day, when Dale's supervisor was sitting in it, the chair suddenly collapsed beneath him. He ordered a new one on the spot. When it came, Minou took over the old one.

Minou hired himself on at the elevator, wages in milk, please.

Soon he had his own litter box. When I asked Dale who paid for the litter, he said it was written off as an expense under mouse control. I had a good laugh over that, but Dale didn't see the humour.

"It's perfectly legitimate," he sniffed. "My supervisor knows about it and lots of elevators do it." So there.

We had a good four years in Provost, but all around us wooden elevators were closing rapidly. New, high throughput terminals were rising to replace them. As he engaged in fact-finding missions, and the selection of a site for the new concrete facility scheduled for Provost, Dale knew his time was limited. The company was leaning towards younger, university-educated men to run the new terminals. As he approached fifty, time was against him. Dale oversaw much of the building of the new plant. Just before it opened, he transferred to Coronation, a smaller, older station slated for closure within the year. We were just buying time. We needed to assess our plans for the future. In December 2000, Dale took a severance package and left the company.

The day he turned over his keys, the worst blizzard of the winter was raging outside. He came home and collected me from an empty house. The car, the van, and the cats were packed. The movers had left the night before. Nineteen-year-old Brandi was waiting for us in Lethbridge. We took a deep breath and together made the decision to head out into whiteout conditions. No regrets. We'd had a wonderful twenty years in the elevator system. We couldn't see where we were going, but we were heading for sunnier skies. With or without the elevators, we figured the next twenty years would be just as good.

# Glossary

**annex**—Structure containing additional storage bins annexed onto the side of an elevator. Some were of cribbed construction, while others were balloon framed. Balloon annexes were distinguishable by the horizontal bands embracing their exterior. These bands anchored tie rods that criss-crossed the interior to support the bins.

**Agro facility**—Separate building from which services other than grain handling were provided. Product sales (chemical, fertilizer, seed, twine) and soil-sampling services are just two examples.

**chop**—Grain that is finely ground into feed for animals.

**cupola**—Smaller, house-shaped edifice that forms the peak of an elevator. It caps the top of the leg mechanism, protecting it from the elements.

**dockage**—The percentage of foreign matter (weed seed, dirt, stones, and other foreign materials) found in a load of grain and for which a farmer is docked. Determined by running a representative sample from the load through a series of screens.

**elevator row**—Term given to the road along which a town's elevators are situated. Once a feature of nearly every prairie community, elevator rows are now rare.

**fanning mill**—Large fan that blows chaff, small seeds, and other lightweight impurities from grain during the cleaning process.

**grade**—Standards set by the Canadian Grain Commission for grain quality. For example, the baseline grade for wheat is #1 Canada Western Red Spring (CWRS). If de-grading factors are present, such as high moisture content, exposure to frost, poor colour, sprouting, et cetera, the wheat will be assigned #2 CWRS, #3 CWRS, or Canada Feed, all progressively inferior grades.

**high throughput facility**—New trend in elevator construction, consisting of a series of concrete or steel silo-like structures with an exposed leg at the top. Computer controlled, they are able to both receive and ship grain simultaneously, loading 50 to 110 railcars in eight to twelve hours.

**permit books**—Issued for each farmer so that the elevator agent can keep track of his/her grain deliveries for the Canadian Wheat Board. Before the introduction of computers, permit books were the agent's primary record of each farmer's grain handling activity.

**quota**—A call for farmers to deliver a specific amount, type, and grade of grain to the elevator. The Canadian Wheat Board issues quotas in response to export sales of Canadian grain.

**scalper**—Device used for scalping grain (removing beards from barley and kernels from unthreshed heads of wheat).

**screenings**—Everything that is left over after cleaning grain.

**spot**—The act of positioning cars in the proper location relative to the elevator for loading.

Both the track and the elevators are now gone.
East of Whitelaw, AB, 1980s.
PHOTOGRAPHER/CONTRIBUTOR: TEENA FENIAK

# Acknowledgments

A friend of mine once wrote: "I'm amazed by the amount of work other people contribute to a book, and the book still has just the author's name on it." I couldn't agree more. Assisting with this book were Clarence Cluff, my primary research source for the early years, and Dale McLachlan, my 'technical expert.' Debbie Rea was a willing Saskatchewan source. Chris Stackhouse went to special lengths with photographs. Dr. Lynne Van Luven was a sensitive and supportive editor, and Christine Savage a competent copy editor. Warm thanks to Sharon Butala for the introduction, and to Ruth Linka, Rebecca Whitney, Darcia Dahl, and all the wonderful staff and directors of NeWest Press. I am particularly indebted to the many who expressed support of the project, even though there were some whose stories or pictures I couldn't use. Of special note are Judy Larmour, Carol Berger, Lynda Swanson, Rhoda Olsen, Inez Wickstrom, Addie Levy and Trev. Quinn. I also wish to acknowledge the employees of Ask A Question, the outstanding virtual reference service of The Alberta Library. When time was short and resources uncertain they consistently went beyond the call, promptly finding me the answers and direction I needed.

Born and raised in rural Alberta, ELIZABETH MCLACHLAN
was inspired to write the personal history of Prairie grain elevators
because of her twenty years as the wife of an elevator agent in
Alberta's rural communities. McLachlan's two previous Prairie his-
tories, *With Unfailing Dedication: Rural Teachers in the War Years*
and *With Unshakeable Persistence: Rural Teachers of the Depression
Era* were also published by NeWest Press. In all three books
McLachlan uses her own experience and the rich experiences of the
communities around her to explore intriguing and important times
in Canadian history. McLachlan now makes her home in
Claresholm, Alberta.